Finite Difference Computing with PDEs

Finite Difference Computing with PDEs

Andy Flair

STATES
ACADEMIC PRESS
www.statesacademicpress.com

Published by States Academic Press
109 South 5th Street,
Brooklyn, NY 11249, USA
www.statesacademicpress.com

Finite Difference Computing with PDEs
Andy Flair

International Standard Book Number: 978-1-63989-203-7 (Hardback)

Cataloging-in-Publication Data

Finite difference computing with PDEs / Andy Flair.
p. cm.
Includes bibliographical references and index.
ISBN 978-1-63989-203-7
1. Differential equations, Partial--Data processing. 2. Finite differences.
3. Computer programming--Mathematics. 4. Differential equations,
Partial--Numerical solutions. I. Flair, Andy.

QA377 .F56 2022
515.353--dc23

Contents

Preface

An equation that relates to one or more functions with its derivatives is called a differential equation. A partial differential equation (PDE) is a type of differential equation, in which the equation consists of unknown multi variables with their partial derivatives. This is a special case of an ordinary differential equation. There is a large amount of modern mathematical and scientific research on methods for numerically approximating solutions of particular PDEs using computers. These are numerical methods for approximating the solutions to differential equations using finite difference equations to approximate derivatives. Finite difference computing is one of the most widely used methods for solving PDEs. This book outlines the processes and applications of finite difference computing with PDEs in detail. The topics included herein on finite difference computing with PDEs are of utmost significance and bound to provide incredible insights to readers. The book is appropriate for students seeking detailed information in this area as well as for experts.

Various studies have approached the subject by analyzing it with a single perspective, but the present book provides diverse methodologies and techniques to address this field. This book contains theories and applications needed for understanding the subject from different perspectives. The aim is to keep the readers informed about the progresses in the field; therefore, the contributions were carefully examined to compile novel researches by specialists from across the globe.

Indeed, the job of the editor is the most crucial and challenging in compiling all chapters into a single book. In the end, I would extend my sincere thanks to the chapter authors for their profound work. I am also thankful for the support provided by my family and colleagues during the compilation of this book.

Andy Flair

1

Vibration ODEs: Schemes, Methods and Models

Vibration problems lead to differential equations with solutions that oscillate in time, typically in a damped or undamped sinusoidal fashion. Such solutions put certain demands on the numerical methods compared to other phenomena whose solutions are monotone or very smooth. Both the frequency and amplitude of the oscillations need to be accurately handled by the numerical schemes. The forthcoming text presents a range of different methods, from classical ones (Runge-Kutta and midpoint/Crank-Nicolson methods), to more modern and popular symplectic (geometric) integration schemes (Leapfrog, Euler-Cromer, and Störmer-Verlet methods), but with a clear emphasis on the latter. Vibration problems occur throughout mechanics and physics, but the methods discussed in this text are also fundamental for constructing successful algorithms for partial differential equations of wave nature in multiple spatial dimensions.

1.1 Finite Difference Discretization

Many of the numerical challenges faced when computing oscillatory solutions to ODEs and PDEs can be captured by the very simple ODE $u'' + u = 0$. This ODE is thus chosen as our starting point for method development, implementation, and analysis.

1.1.1 A Basic Model for Vibrations

The simplest model of a vibrating mechanical system has the following form:

$$u'' + \omega^2 u = 0, \quad u(0) = I,\ u'(0) = 0,\ t \in (0, T]\,. \tag{1.1}$$

Here, ω and I are given constants. Section 1.12.1 derives (1.1) from physical principles and explains what the constants mean.

The exact solution of (1.1) is

$$u(t) = I\cos(\omega t)\,. \tag{1.2}$$

That is, u oscillates with constant amplitude I and angular frequency ω. The corresponding period of oscillations (i.e., the time between two neighboring peaks in the cosine function) is $P = 2\pi/\omega$. The number of periods per second is $f = \omega/(2\pi)$ and measured in the unit Hz. Both f and ω are referred to as frequency, but ω is more precisely named *angular frequency*, measured in rad/s.

In vibrating mechanical systems modeled by (1.1), $u(t)$ very often represents a position or a displacement of a particular point in the system. The derivative $u'(t)$ then has the interpretation of velocity, and $u''(t)$ is the associated acceleration. The model (1.1) is not only applicable to vibrating mechanical systems, but also to oscillations in electrical circuits.

1.1.2 A Centered Finite Difference Scheme

To formulate a finite difference method for the model problem (1.1), we follow the four steps explained in Section 1.1.2 in [9].

Step 1: Discretizing the domain The domain is discretized by introducing a uniformly partitioned time mesh. The points in the mesh are $t_n = n\Delta t$, $n = 0, 1, \ldots, N_t$, where $\Delta t = T/N_t$ is the constant length of the time steps. We introduce a mesh function u^n for $n = 0, 1, \ldots, N_t$, which approximates the exact solution at the mesh points. (Note that $n = 0$ is the known initial condition, so u^n is identical to the mathematical u at this point.) The mesh function u^n will be computed from algebraic equations derived from the differential equation problem.

Step 2: Fulfilling the equation at discrete time points The ODE is to be satisfied at each mesh point where the solution must be found:

$$u''(t_n) + \omega^2 u(t_n) = 0, \quad n = 1, \ldots, N_t\,. \tag{1.3}$$

Step 3: Replacing derivatives by finite differences The derivative $u''(t_n)$ is to be replaced by a finite difference approximation. A common second-order accurate approximation to the second-order derivative is

$$u''(t_n) \approx \frac{u^{n+1} - 2u^n + u^{n-1}}{\Delta t^2}\,. \tag{1.4}$$

Inserting (1.4) in (1.3) yields

$$\frac{u^{n+1} - 2u^n + u^{n-1}}{\Delta t^2} = -\omega^2 u^n\,. \tag{1.5}$$

We also need to replace the derivative in the initial condition by a finite difference. Here we choose a centered difference, whose accuracy is similar to the centered difference we used for u'':

$$\frac{u^1 - u^{-1}}{2\Delta t} = 0\,. \tag{1.6}$$

Step 4: Formulating a recursive algorithm To formulate the computational algorithm, we assume that we have already computed u^{n-1} and u^n, such that u^{n+1} is the unknown value to be solved for:

$$u^{n+1} = 2u^n - u^{n-1} - \Delta t^2\omega^2 u^n\,. \tag{1.7}$$

The computational algorithm is simply to apply (1.7) successively for $n = 1, 2, \ldots, N_t - 1$. This numerical scheme sometimes goes under the name Störmer's method, Verlet integration[1], or the Leapfrog method (one should note that Leapfrog is used for many quite different methods for quite different differential equations!).

Computing the first step We observe that (1.7) cannot be used for $n = 0$ since the computation of u^1 then involves the undefined value u^{-1} at $t = -\Delta t$. The discretization of the initial condition then comes to our rescue: (1.6) implies $u^{-1} = u^1$ and this relation can be combined with (1.7) for $n = 0$ to yield a value for u^1:

$$u^1 = 2u^0 - u^1 - \Delta t^2\omega^2 u^0,$$

which reduces to

$$u^1 = u^0 - \frac{1}{2}\Delta t^2\omega^2 u^0\,. \tag{1.8}$$

Exercise 1.5 asks you to perform an alternative derivation and also to generalize the initial condition to $u'(0) = V \neq 0$.

The computational algorithm The steps for solving (1.1) become

1. $u^0 = I$
2. compute u^1 from (1.8)
3. for $n = 1, 2, \ldots, N_t - 1$: compute u^{n+1} from (1.7)

The algorithm is more precisely expressed directly in Python:

```
t = linspace(0, T, Nt+1)  # mesh points in time
dt = t[1] - t[0]          # constant time step
u = zeros(Nt+1)           # solution

u[0] = I
u[1] = u[0] - 0.5*dt**2*w**2*u[0]
for n in range(1, Nt):
    u[n+1] = 2*u[n] - u[n-1] - dt**2*w**2*u[n]
```

Remark on using `w` for ω in computer code

In the code, we use `w` as the symbol for ω. The reason is that the authors prefer `w` for readability and comparison with the mathematical ω instead of the full word `omega` as variable name.

[1] http://en.wikipedia.org/wiki/Verlet_integration

Operator notation We may write the scheme using a compact difference notation listed in Appendix A.1 (see also Section 1.1.8 in [9]). The difference (1.4) has the operator notation $[D_t D_t u]^n$ such that we can write:

$$[D_t D_t u + \omega^2 u = 0]^n \,. \tag{1.9}$$

Note that $[D_t D_t u]^n$ means applying a central difference with step $\Delta t/2$ twice:

$$[D_t(D_t u)]^n = \frac{[D_t u]^{n+\frac{1}{2}} - [D_t u]^{n-\frac{1}{2}}}{\Delta t}$$

which is written out as

$$\frac{1}{\Delta t}\left(\frac{u^{n+1}-u^n}{\Delta t} - \frac{u^n - u^{n-1}}{\Delta t}\right) = \frac{u^{n+1} - 2u^n + u^{n-1}}{\Delta t^2}\,.$$

The discretization of initial conditions can in the operator notation be expressed as

$$[u = I]^0, \quad [D_{2t} u = 0]^0, \tag{1.10}$$

where the operator $[D_{2t} u]^n$ is defined as

$$[D_{2t} u]^n = \frac{u^{n+1} - u^{n-1}}{2\Delta t}\,. \tag{1.11}$$

1.2 Implementation

1.2.1 Making a Solver Function

The algorithm from the previous section is readily translated to a complete Python function for computing and returning $u^0, u^1, \ldots, u^{N_t}$ and $t_0, t_1, \ldots, t_{N_t}$, given the input I, ω, Δt, and T:

```
import numpy as np
import matplotlib.pyplot as plt

def solver(I, w, dt, T):
    """
    Solve u'' + w**2*u = 0 for t in (0,T], u(0)=I and u'(0)=0,
    by a central finite difference method with time step dt.
    """
    dt = float(dt)
    Nt = int(round(T/dt))
    u = np.zeros(Nt+1)
    t = np.linspace(0, Nt*dt, Nt+1)

    u[0] = I
    u[1] = u[0] - 0.5*dt**2*w**2*u[0]
    for n in range(1, Nt):
        u[n+1] = 2*u[n] - u[n-1] - dt**2*w**2*u[n]
    return u, t
```

We have imported numpy and matplotlib under the names np and plt, respectively, as this is very common in the Python scientific computing community and a good programming habit (since we explicitly see where the different functions come from). An alternative is to do from numpy import * and a similar “import all” for Matplotlib to avoid the np and plt prefixes and make the code as close as possible to MATLAB. (See Section 5.1.4 in [9] for a discussion of the two types of import in Python.)

A function for plotting the numerical and the exact solution is also convenient to have:

```
def u_exact(t, I, w):
    return I*np.cos(w*t)

def visualize(u, t, I, w):
    plt.plot(t, u, 'r--o')
    t_fine = np.linspace(0, t[-1], 1001)  # very fine mesh for u_e
    u_e = u_exact(t_fine, I, w)
    plt.hold('on')
    plt.plot(t_fine, u_e, 'b-')
    plt.legend(['numerical', 'exact'], loc='upper left')
    plt.xlabel('t')
    plt.ylabel('u')
    dt = t[1] - t[0]
    plt.title('dt=%g' % dt)
    umin = 1.2*u.min();  umax = -umin
    plt.axis([t[0], t[-1], umin, umax])
    plt.savefig('tmp1.png');  plt.savefig('tmp1.pdf')
```

A corresponding main program calling these functions to simulate a given number of periods (num_periods) may take the form

```
I = 1
w = 2*pi
dt = 0.05
num_periods = 5
P = 2*pi/w    #  one period
T = P*num_periods
u, t = solver(I, w, dt, T)
visualize(u, t, I, w, dt)
```

Adjusting some of the input parameters via the command line can be handy. Here is a code segment using the ArgumentParser tool in the argparse module to define option value (-option value) pairs on the command line:

```
import argparse
parser = argparse.ArgumentParser()
parser.add_argument('--I', type=float, default=1.0)
parser.add_argument('--w', type=float, default=2*pi)
parser.add_argument('--dt', type=float, default=0.05)
parser.add_argument('--num_periods', type=int, default=5)
a = parser.parse_args()
I, w, dt, num_periods = a.I, a.w, a.dt, a.num_periods
```

Such parsing of the command line is explained in more detail in Section 5.2.3 in [9].

A typical execution goes like

Terminal

```
Terminal> python vib_undamped.py --num_periods 20 --dt 0.1
```

Computing u' In mechanical vibration applications one is often interested in computing the velocity $v(t) = u'(t)$ after $u(t)$ has been computed. This can be done by a central difference,

$$v(t_n) = u'(t_n) \approx v^n = \frac{u^{n+1} - u^{n-1}}{2\Delta t} = [D_{2t}u]^n\,. \tag{1.12}$$

This formula applies for all inner mesh points, $n = 1, \ldots, N_t - 1$. For $n = 0$, $v(0)$ is given by the initial condition on $u'(0)$, and for $n = N_t$ we can use a one-sided, backward difference:

$$v^n = [D_t^- u]^n = \frac{u^n - u^{n-1}}{\Delta t}\,.$$

Typical (scalar) code is

```
v = np.zeros_like(u)  # or v = np.zeros(len(u))
# Use central difference for internal points
for i in range(1, len(u)-1):
    v[i] = (u[i+1] - u[i-1])/(2*dt)
# Use initial condition for u'(0) when i=0
v[0] = 0
# Use backward difference at the final mesh point
v[-1] = (u[-1] - u[-2])/dt
```

Since the loop is slow for large N_t, we can get rid of the loop by vectorizing the central difference. The above code segment goes as follows in its vectorized version (see Problem 1.2 in [9] for explanation of details):

```
v = np.zeros_like(u)
v[1:-1] = (u[2:] - u[:-2])/(2*dt)  # central difference
v[0] = 0                           # boundary condition u'(0)
v[-1] = (u[-1] - u[-2])/dt         # backward difference
```

1.2.2 Verification

Manual calculation The simplest type of verification, which is also instructive for understanding the algorithm, is to compute u^1, u^2, and u^3 with the aid of a calculator and make a function for comparing these results with those from the `solver` function. The `test_three_steps` function in the file `vib_undamped.py` shows the details of how we use the hand calculations to test the code:

```
def test_three_steps():
    from math import pi
    I = 1;  w = 2*pi;  dt = 0.1;  T = 1
    u_by_hand = np.array([1.000000000000000,
                          0.802607911978213,
                          0.288358920740053])
    u, t = solver(I, w, dt, T)
    diff = np.abs(u_by_hand - u[:3]).max()
    tol = 1E-14
    assert diff < tol
```

This function is a proper *test function*, compliant with the pytest and nose testing framework for Python code, because

- the function name begins with `test_`
- the function takes no arguments
- the test is formulated as a boolean condition and executed by `assert`

We shall in this book implement all software verification via such proper test functions, also known as unit testing. See Section 5.3.2 in [9] for more details on how to construct test functions and utilize nose or pytest for automatic execution of tests. Our recommendation is to use pytest. With this choice, you can run all test functions in `vib_undamped.py` by

Terminal

```
Terminal> py.test -s -v vib_undamped.py
============================= test session starts ======...
platform linux2 -- Python 2.7.9 -- ...
collected 2 items

vib_undamped.py::test_three_steps PASSED
vib_undamped.py::test_convergence_rates PASSED

========================== 2 passed in 0.19 seconds ===...
```

Testing very simple polynomial solutions Constructing test problems where the exact solution is constant or linear helps initial debugging and verification as one expects any reasonable numerical method to reproduce such solutions to machine precision. Second-order accurate methods will often also reproduce a quadratic solution. Here $[D_tD_tt^2]^n = 2$, which is the exact result. A solution $u = t^2$ leads to $u'' + \omega^2 u = 2 + (\omega t)^2 \neq 0$. We must therefore add a source in the equation: $u'' + \omega^2 u = f$ to allow a solution $u = t^2$ for $f = 2 + (\omega t)^2$. By simple insertion we can show that the mesh function $u^n = t_n^2$ is also a solution of the discrete equations. Problem 1.1 asks you to carry out all details to show that linear and quadratic solutions are solutions of the discrete equations. Such results are very useful for debugging and verification. You are strongly encouraged to do this problem now!

Checking convergence rates Empirical computation of convergence rates yields a good method for verification. The method and its computational details are explained in detail in Section 3.1.6 in [9]. Readers not familiar with the concept should look up this reference before proceeding.

In the present problem, computing convergence rates means that we must

- perform m simulations, halving the time steps as: $\Delta t_i = 2^{-i}\Delta t_0$, $i = 1, \ldots, m-1$, and Δt_i is the time step used in simulation i;
- compute the L^2 norm of the error, $E_i = \sqrt{\Delta t_i \sum_{n=0}^{N_t-1}(u^n - u_e(t_n))^2}$ in each case;
- estimate the convergence rates r_i based on two consecutive experiments $(\Delta t_{i-1}, E_{i-1})$ and $(\Delta t_i, E_i)$, assuming $E_i = C(\Delta t_i)^r$ and $E_{i-1} = C(\Delta t_{i-1})^r$, where C is a constant. From these equations it follows that $r = \ln(E_{i-1}/E_i)/\ln(\Delta t_{i-1}/\Delta t_i)$. Since this r will vary with i, we equip it with an index and call it r_{i-1}, where i runs from 1 to $m-1$.

The computed rates $r_0, r_1, \ldots, r_{m-2}$ hopefully converge to the number 2 in the present problem, because theory (from Sect. 1.4) shows that the error of the numerical method we use behaves like Δt^2. The convergence of the sequence $r_0, r_1, \ldots, r_{m-2}$ demands that the time steps Δt_i are sufficiently small for the error model $E_i = C(\Delta t_i)^r$ to be valid.

All the implementational details of computing the sequence $r_0, r_1, \ldots, r_{m-2}$ appear below.

```
def convergence_rates(m, solver_function, num_periods=8):
    """
    Return m-1 empirical estimates of the convergence rate
    based on m simulations, where the time step is halved
    for each simulation.
    solver_function(I, w, dt, T) solves each problem, where T
    is based on simulation for num_periods periods.
    """
    from math import pi
    w = 0.35; I = 0.3          # just chosen values
    P = 2*pi/w                 # period
    dt = P/30                  # 30 time step per period 2*pi/w
    T = P*num_periods

    dt_values = []
    E_values = []
    for i in range(m):
        u, t = solver_function(I, w, dt, T)
        u_e = u_exact(t, I, w)
        E = np.sqrt(dt*np.sum((u_e-u)**2))
        dt_values.append(dt)
        E_values.append(E)
        dt = dt/2

    r = [np.log(E_values[i-1]/E_values[i])/
         np.log(dt_values[i-1]/dt_values[i])
         for i in range(1, m, 1)]
    return r, E_values, dt_values
```

The error analysis in Sect. 1.4 is quite detailed and suggests that $r = 2$. It is also a intuitively reasonable result, since we used a second-order accurate finite difference approximation $[D_tD_tu]^n$ to the ODE and a second-order accurate finite difference formula for the initial condition for u'.

In the present problem, when Δt_0 corresponds to 30 time steps per period, the returned `r` list has all its values equal to 2.00 (if rounded to two decimals). This amazingly accurate result means that all Δt_i values are well into the asymptotic regime where the error model $E_i = C(\Delta t_i)^r$ is valid.

We can now construct a proper test function that computes convergence rates and checks that the final (and usually the best) estimate is sufficiently close to 2. Here, a rough tolerance of 0.1 is enough. Later, we will argue for an improvement by adjusting omega and include also that case in our test function here. The unit test goes like

```
def test_convergence_rates():
    r, E, dt = convergence_rates(
        m=5, solver_function=solver, num_periods=8)
    # Accept rate to 1 decimal place
    tol = 0.1
    assert abs(r[-1] - 2.0) < tol
    # Test that adjusted w obtains 4th order convergence
    r, E, dt = convergence_rates(
        m=5, solver_function=solver_adjust_w, num_periods=8)
    print 'adjust w rates:', r
    assert abs(r[-1] - 4.0) < tol
```

The complete code appears in the file `vib_undamped.py`.

Visualizing convergence rates with slope markers Tony S. Yu has written a script `plotslopes.py`[2] that is very useful to indicate the slope of a graph, especially a graph like $\ln E = r \ln \Delta t + \ln C$ arising from the model $E = C\Delta t^r$. A copy of the script resides in the `src/vib`[3] directory. Let us use it to compare the original method for $u'' + \omega^2 u = 0$ with the same method applied to the equation with a modified ω. We make log-log plots of the error versus Δt. For each curve we attach a slope marker using the `slope_marker((x,y), r)` function from `plotslopes.py`, where `(x,y)` is the position of the marker and `r` and the slope $((r, 1))$, here (2,1) and (4,1).

```
def plot_convergence_rates():
    r2, E2, dt2 = convergence_rates(
        m=5, solver_function=solver, num_periods=8)
    plt.loglog(dt2, E2)
    r4, E4, dt4 = convergence_rates(
        m=5, solver_function=solver_adjust_w, num_periods=8)
    plt.loglog(dt4, E4)
    plt.legend(['original scheme', r'adjusted $\omega$'],
               loc='upper left')
    plt.title('Convergence of finite difference methods')
    from plotslopes import slope_marker
    slope_marker((dt2[1], E2[1]), (2,1))
    slope_marker((dt4[1], E4[1]), (4,1))
```

[2] http://goo.gl/A4Utm7

[3] http://tinyurl.com/nu656p2/vib

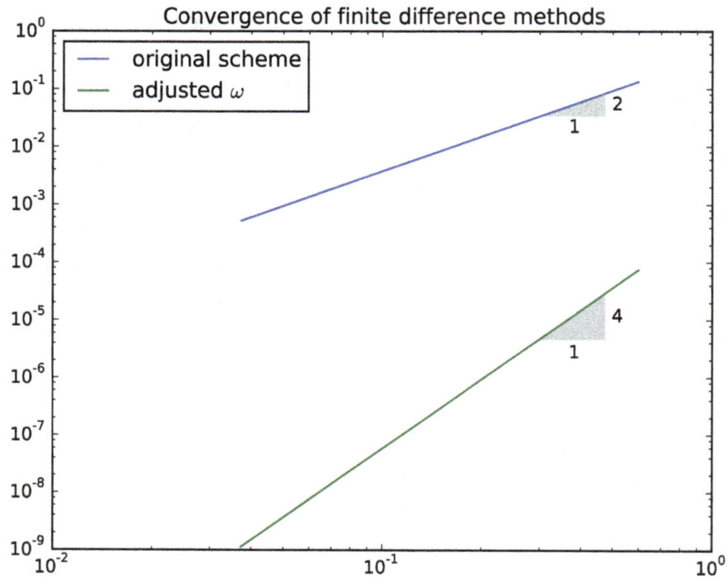

Fig. 1.1 Empirical convergence rate curves with special slope marker

Figure 1.1 displays the two curves with the markers. The match of the curve slope and the marker slope is excellent.

1.2.3 Scaled Model

It is advantageous to use dimensionless variables in simulations, because fewer parameters need to be set. The present problem is made dimensionless by introducing dimensionless variables $\bar{t} = t/t_c$ and $\bar{u} = u/u_c$, where t_c and u_c are characteristic scales for t and u, respectively. We refer to Section 2.2.1 in [11] for all details about this scaling.

The scaled ODE problem reads

$$\frac{u_c}{t_c^2}\frac{d^2\bar{u}}{d\bar{t}^2} + u_c\bar{u} = 0, \quad u_c\bar{u}(0) = I,\ \frac{u_c}{t_c}\frac{d\bar{u}}{d\bar{t}}(0) = 0\,.$$

A common choice is to take t_c as one period of the oscillations, $t_c = 2\pi/w$, and $u_c = I$. This gives the dimensionless model

$$\frac{d^2\bar{u}}{d\bar{t}^2} + 4\pi^2\bar{u} = 0, \quad \bar{u}(0) = 1,\ \bar{u}'(0) = 0\,. \tag{1.13}$$

Observe that there are no physical parameters in (1.13)! We can therefore perform a single numerical simulation $\bar{u}(\bar{t})$ and afterwards recover any $u(t;\omega, I)$ by

$$u(t;\omega, I) = u_c\bar{u}(t/t_c) = I\bar{u}(\omega t/(2\pi))\,.$$

We can easily check this assertion: the solution of the scaled problem is $\bar{u}(\bar{t}) = \cos(2\pi\bar{t})$. The formula for u in terms of $\bar{u}$ gives $u = I\cos(\omega t)$, which is nothing but the solution of the original problem with dimensions.

The scaled model can be run by calling `solver(I=1, w=2*pi, dt, T)`. Each period is now 1 and `T` simply counts the number of periods. Choosing `dt` as `1./M` gives `M` time steps per period.

1.3 Visualization of Long Time Simulations

Figure 1.2 shows a comparison of the exact and numerical solution for the scaled model (1.13) with $\Delta t = 0.1, 0.05$. From the plot we make the following observations:

- The numerical solution seems to have correct amplitude.
- There is an angular frequency error which is reduced by decreasing the time step.
- The total angular frequency error grows with time.

By angular frequency error we mean that the numerical angular frequency differs from the exact ω. This is evident by looking at the peaks of the numerical solution: these have incorrect positions compared with the peaks of the exact cosine solution. The effect can be mathematically expressed by writing the numerical solution as $I \cos \tilde{\omega} t$, where $\tilde{\omega}$ is not exactly equal to ω. Later, we shall mathematically quantify this numerical angular frequency $\tilde{\omega}$.

1.3.1 Using a Moving Plot Window

In vibration problems it is often of interest to investigate the system's behavior over long time intervals. Errors in the angular frequency accumulate and become more visible as time grows. We can investigate long time series by introducing a moving plot window that can move along with the p most recently computed periods of the solution. The SciTools[4] package contains a convenient tool for this: `MovingPlotWindow`. Typing `pydoc scitools.MovingPlotWindow` shows a demo and a description of its use. The function below utilizes the moving plot

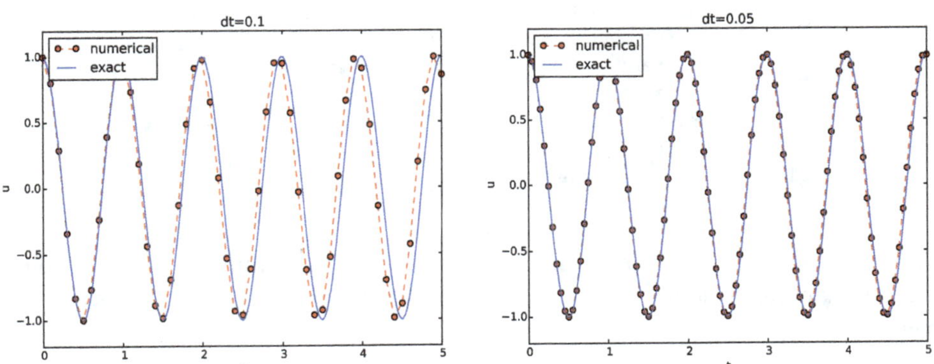

Fig. 1.2 Effect of halving the time step

[4] https://github.com/hplgit/scitools

window and is in fact called by the `main` function in the `vib_undamped` module if the number of periods in the simulation exceeds 10.

```
def visualize_front(u, t, I, w, savefig=False, skip_frames=1):
    """
    Visualize u and the exact solution vs t, using a
    moving plot window and continuous drawing of the
    curves as they evolve in time.
    Makes it easy to plot very long time series.
    Plots are saved to files if savefig is True.
    Only each skip_frames-th plot is saved (e.g., if
    skip_frame=10, only each 10th plot is saved to file;
    this is convenient if plot files corresponding to
    different time steps are to be compared).
    """
    import scitools.std as st
    from scitools.MovingPlotWindow import MovingPlotWindow
    from math import pi

    # Remove all old plot files tmp_*.png
    import glob, os
    for filename in glob.glob('tmp_*.png'):
        os.remove(filename)

    P = 2*pi/w  # one period
    umin = 1.2*u.min();  umax = -umin
    dt = t[1] - t[0]
    plot_manager = MovingPlotWindow(
        window_width=8*P,
        dt=dt,
        yaxis=[umin, umax],
        mode='continuous drawing')
    frame_counter = 0
    for n in range(1,len(u)):
        if plot_manager.plot(n):
            s = plot_manager.first_index_in_plot
            st.plot(t[s:n+1], u[s:n+1], 'r-1',
                    t[s:n+1], I*cos(w*t)[s:n+1], 'b-1',
                    title='t=%6.3f' % t[n],
                    axis=plot_manager.axis(),
                    show=not savefig) # drop window if savefig
            if savefig and n % skip_frames == 0:
                filename = 'tmp_%04d.png' % frame_counter
                st.savefig(filename)
                print 'making plot file', filename, 'at t=%g' % t[n]
                frame_counter += 1
        plot_manager.update(n)
```

We run the scaled problem (the default values for the command-line arguments `-I` and `-w` correspond to the scaled problem) for 40 periods with 20 time steps per period:

Terminal

```
Terminal> python vib_undamped.py --dt 0.05 --num_periods 40
```

The moving plot window is invoked, and we can follow the numerical and exact solutions as time progresses. From this demo we see that the angular frequency error is small in the beginning, and that it becomes more prominent with time. A new run with $\Delta t = 0.1$ (i.e., only 10 time steps per period) clearly shows that the phase errors become significant even earlier in the time series, deteriorating the solution further.

1.3.2 Making Animations

Producing standard video formats The `visualize_front` function stores all the plots in files whose names are numbered: `tmp_0000.png`, `tmp_0001.png`, `tmp_0002.png`, and so on. From these files we may make a movie. The Flash format is popular,

Terminal
```
Terminal> ffmpeg -r 25 -i tmp_%04d.png -c:v flv movie.flv
```

The `ffmpeg` program can be replaced by the `avconv` program in the above command if desired (but at the time of this writing it seems to be more momentum in the `ffmpeg` project). The `-r` option should come first and describes the number of frames per second in the movie (even if we would like to have slow movies, keep this number as large as 25, otherwise files are skipped from the movie). The `-i` option describes the name of the plot files. Other formats can be generated by changing the video codec and equipping the video file with the right extension:

Format	Codec and filename
Flash	`-c:v flv movie.flv`
MP4	`-c:v libx264 movie.mp4`
WebM	`-c:v libvpx movie.webm`
Ogg	`-c:v libtheora movie.ogg`

The video file can be played by some video player like `vlc`, `mplayer`, `gxine`, or `totem`, e.g.,

Terminal
```
Terminal> vlc movie.webm
```

A web page can also be used to play the movie. Today's standard is to use the HTML5 `video` tag:

```
<video autoplay loop controls
       width='640' height='365' preload='none'>
<source src='movie.webm'  type='video/webm; codecs="vp8, vorbis"'>
</video>
```

Modern browsers do not support all of the video formats. MP4 is needed to successfully play the videos on Apple devices that use the Safari browser. WebM is the preferred format for Chrome, Opera, Firefox, and Internet Explorer v9+. Flash was a popular format, but older browsers that required Flash can play MP4. All browsers that work with Ogg can also work with WebM. This means that to have a video work in all browsers, the video should be available in the MP4 and WebM formats. The proper HTML code reads

```
<video autoplay loop controls
       width='640' height='365' preload='none'>
<source src='movie.mp4'   type='video/mp4;
 codecs="avc1.42E01E, mp4a.40.2"'>
<source src='movie.webm'  type='video/webm;
 codecs="vp8, vorbis"'>
</video>
```

The MP4 format should appear first to ensure that Apple devices will load the video correctly.

Caution: number the plot files correctly

To ensure that the individual plot frames are shown in correct order, it is important to number the files with zero-padded numbers (0000, 0001, 0002, etc.). The printf format `%04d` specifies an integer in a field of width 4, padded with zeros from the left. A simple Unix wildcard file specification like `tmp_*.png` will then list the frames in the right order. If the numbers in the filenames were not zero-padded, the frame `tmp_11.png` would appear before `tmp_2.png` in the movie.

Playing PNG files in a web browser The `scitools movie` command can create a movie player for a set of PNG files such that a web browser can be used to watch the movie. This interface has the advantage that the speed of the movie can easily be controlled, a feature that scientists often appreciate. The command for creating an HTML with a player for a set of PNG files `tmp_*.png` goes like

```
Terminal> scitools movie output_file=vib.html fps=4 tmp_*.png
```

The `fps` argument controls the speed of the movie ("frames per second").

To watch the movie, load the video file `vib.html` into some browser, e.g.,

```
Terminal> google-chrome vib.html  # invoke web page
```

Click on `Start movie` to see the result. Moving this movie to some other place requires moving `vib.html` *and all the PNG files* `tmp_*.png`:

Terminal
```
Terminal> mkdir vib_dt0.1
Terminal> mv tmp_*.png vib_dt0.1
Terminal> mv vib.html vib_dt0.1/index.html
```

Making animated GIF files The `convert` program from the ImageMagick software suite can be used to produce animated GIF files from a set of PNG files:

Terminal
```
Terminal> convert -delay 25 tmp_vib*.png tmp_vib.gif
```

The `-delay` option needs an argument of the delay between each frame, measured in 1/100 s, so 4 frames/s here gives 25/100 s delay. Note, however, that in this particular example with $\Delta t = 0.05$ and 40 periods, making an animated GIF file out of the large number of PNG files is a very heavy process and not considered feasible. Animated GIFs are best suited for animations with not so many frames and where you want to see each frame and play them slowly.

1.3.3 Using Bokeh to Compare Graphs

Instead of a moving plot frame, one can use tools that allow panning by the mouse. For example, we can show four periods of several signals in several plots and then scroll with the mouse through the rest of the simulation *simultaneously* in all the plot windows. The Bokeh[5] plotting library offers such tools, but the plots must be displayed in a web browser. The documentation of Bokeh is excellent, so here we just show how the library can be used to compare a set of u curves corresponding to long time simulations. (By the way, the guidance to correct pronunciation of Bokeh in the documentation[6] and on Wikipedia[7] is not directly compatible with a YouTube video[8] ...).

Imagine we have performed experiments for a set of Δt values. We want each curve, together with the exact solution, to appear in a plot, and then arrange all plots in a grid-like fashion:

[5] http://bokeh.pydata.org/en/latest
[6] http://bokeh.pydata.org/en/0.10.0/docs/faq.html#how-do-you-pronounce-bokeh
[7] https://en.wikipedia.org/wiki/Bokeh
[8] https://www.youtube.com/watch?v=OR8HSHevQTM

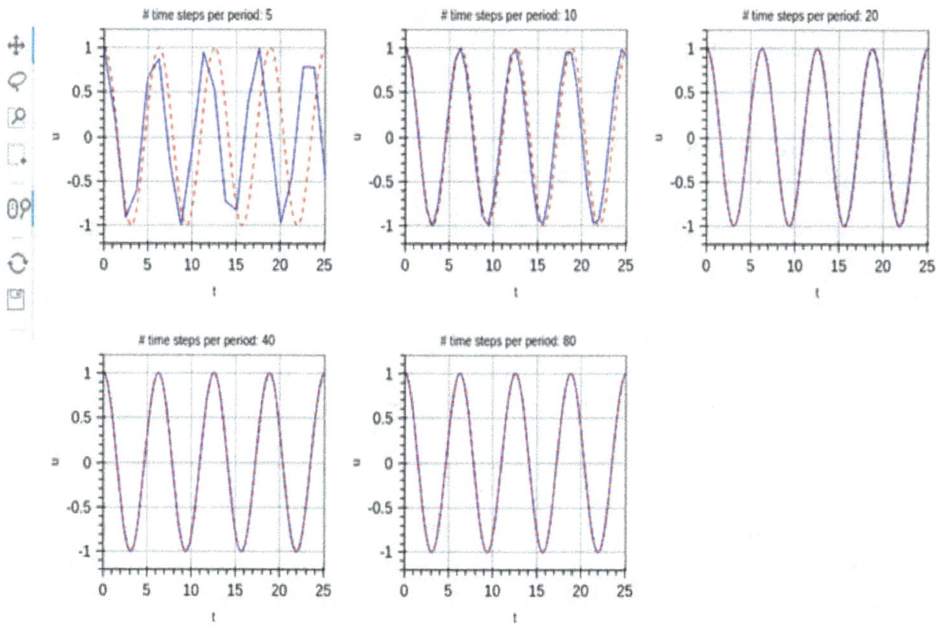

Furthermore, we want the axes to couple such that if we move into the future in one plot, all the other plots follows (note the displaced t axes!):

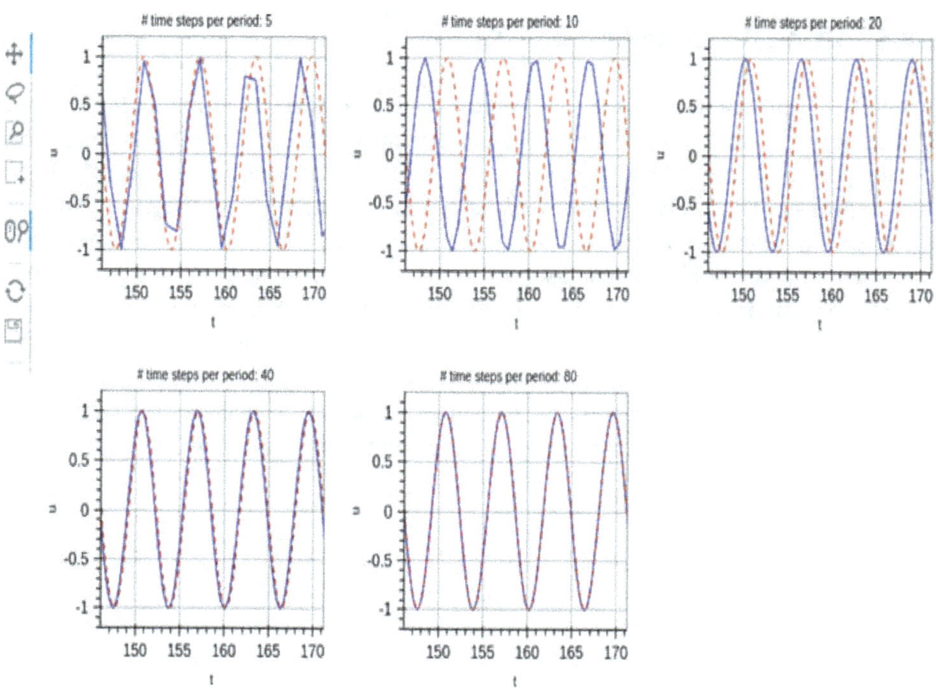

A function for creating a Bokeh plot, given a list of `u` arrays and corresponding `t` arrays, is implemented below. The code combines data from different simulations, described compactly in a list of strings `legends`.

```
def bokeh_plot(u, t, legends, I, w, t_range, filename):
    """
    Make plots for u vs t using the Bokeh library.
    u and t are lists (several experiments can be compared).
    legens contain legend strings for the various u,t pairs.
    """
    if not isinstance(u, (list,tuple)):
        u = [u]  # wrap in list
    if not isinstance(t, (list,tuple)):
        t = [t]  # wrap in list
    if not isinstance(legends, (list,tuple)):
        legends = [legends]  # wrap in list

    import bokeh.plotting as plt
    plt.output_file(filename, mode='cdn', title='Comparison')
    # Assume that all t arrays have the same range
    t_fine = np.linspace(0, t[0][-1], 1001)  # fine mesh for u_e
    tools = 'pan,wheel_zoom,box_zoom,reset,'\
            'save,box_select,lasso_select'
    u_range = [-1.2*I, 1.2*I]
    font_size = '8pt'
    p = []  # list of plot objects
    # Make the first figure
    p_ = plt.figure(
        width=300, plot_height=250, title=legends[0],
        x_axis_label='t', y_axis_label='u',
        x_range=t_range, y_range=u_range, tools=tools,
        title_text_font_size=font_size)
    p_.xaxis.axis_label_text_font_size=font_size
    p_.yaxis.axis_label_text_font_size=font_size
    p_.line(t[0], u[0], line_color='blue')
    # Add exact solution
    u_e = u_exact(t_fine, I, w)
    p_.line(t_fine, u_e, line_color='red', line_dash='4 4')
    p.append(p_)
    # Make the rest of the figures and attach their axes to
    # the first figure's axes
    for i in range(1, len(t)):
        p_ = plt.figure(
            width=300, plot_height=250, title=legends[i],
            x_axis_label='t', y_axis_label='u',
            x_range=p[0].x_range, y_range=p[0].y_range, tools=tools,
            title_text_font_size=font_size)
        p_.xaxis.axis_label_text_font_size = font_size
        p_.yaxis.axis_label_text_font_size = font_size
        p_.line(t[i], u[i], line_color='blue')
        p_.line(t_fine, u_e, line_color='red', line_dash='4 4')
        p.append(p_)

    # Arrange all plots in a grid with 3 plots per row
    grid = [[]]
    for i, p_ in enumerate(p):
        grid[-1].append(p_)
        if (i+1) % 3 == 0:
            # New row
            grid.append([])
    plot = plt.gridplot(grid, toolbar_location='left')
    plt.save(plot)
    plt.show(plot)
```

A particular example using the `bokeh_plot` function appears below.

```
def demo_bokeh():
    """Solve a scaled ODE u'' + u = 0."""
    from math import pi
    w = 1.0           # Scaled problem (frequency)
    P = 2*np.pi/w     # Period
    num_steps_per_period = [5, 10, 20, 40, 80]
    T = 40*P          # Simulation time: 40 periods
    u = []            # List of numerical solutions
    t = []            # List of corresponding meshes
    legends = []
    for n in num_steps_per_period:
        dt = P/n
        u_, t_ = solver(I=1, w=w, dt=dt, T=T)
        u.append(u_)
        t.append(t_)
        legends.append('# time steps per period: %d' % n)
    bokeh_plot(u, t, legends, I=1, w=w, t_range=[0, 4*P],
               filename='tmp.html')
```

1.3.4 Using a Line-by-Line Ascii Plotter

Plotting functions vertically, line by line, in the terminal window using ascii characters only is a simple, fast, and convenient visualization technique for long time series. Note that the time axis then is positive downwards on the screen, so we can let the solution be visualized "forever". The tool `scitools.avplotter.Plotter` makes it easy to create such plots:

```
def visualize_front_ascii(u, t, I, w, fps=10):
    """
    Plot u and the exact solution vs t line by line in a
    terminal window (only using ascii characters).
    Makes it easy to plot very long time series.
    """
    from scitools.avplotter import Plotter
    import time
    from math import pi
    P = 2*pi/w
    umin = 1.2*u.min();  umax = -umin

    p = Plotter(ymin=umin, ymax=umax, width=60, symbols='+o')
    for n in range(len(u)):
        print p.plot(t[n], u[n], I*cos(w*t[n])), \
              '%.1f' % (t[n]/P)
        time.sleep(1/float(fps))
```

The call `p.plot` returns a line of text, with the t axis marked and a symbol + for the first function (`u`) and `o` for the second function (the exact solution). Here we append to this text a time counter reflecting how many periods the current time point corresponds to. A typical output ($\omega = 2\pi$, $\Delta t = 0.05$) looks like this:

```
                         |                       o+      14.0
                         |                      + o      14.0
                         |                  +    o       14.1
                         |             +     o           14.1
                         |     +        o                14.2
                        +|       o                       14.2
                +        |                               14.2
         +       o       |                               14.3
    +     o              |                               14.4
 +   o                   |                               14.4
+o                       |                               14.5
o +                      |                               14.5
 o    +                  |                               14.6
     o      +            |                               14.6
          o        +     |                               14.7
                 o       | +                             14.7
                         |        +                      14.8
                         |       o       +               14.8
                         |              o     +          14.9
                         |                   o   +       14.9
                         |                       o+      15.0
```

1.3.5 Empirical Analysis of the Solution

For oscillating functions like those in Fig. 1.2 we may compute the amplitude and frequency (or period) empirically. That is, we run through the discrete solution points (t_n, u_n) and find all maxima and minima points. The distance between two consecutive maxima (or minima) points can be used as estimate of the local period, while half the difference between the u value at a maximum and a nearby minimum gives an estimate of the local amplitude.

The local maxima are the points where

$$u^{n-1} < u^n > u^{n+1}, \quad n = 1, \ldots, N_t - 1, \tag{1.14}$$

and the local minima are recognized by

$$u^{n-1} > u^n < u^{n+1}, \quad n = 1, \ldots, N_t - 1\,. \tag{1.15}$$

In computer code this becomes

```
def minmax(t, u):
    minima = []; maxima = []
    for n in range(1, len(u)-1, 1):
        if u[n-1] > u[n] < u[n+1]:
            minima.append((t[n], u[n]))
        if u[n-1] < u[n] > u[n+1]:
            maxima.append((t[n], u[n]))
    return minima, maxima
```

Note that the two returned objects are lists of tuples.

Let (t_i, e_i), $i = 0, \ldots, M-1$, be the sequence of all the M maxima points, where t_i is the time value and e_i the corresponding u value. The local period can be defined as $p_i = t_{i+1} - t_i$. With Python syntax this reads

```
def periods(maxima):
    p = [extrema[n][0] - maxima[n-1][0]
         for n in range(1, len(maxima))]
    return np.array(p)
```

The list p created by a list comprehension is converted to an array since we probably want to compute with it, e.g., find the corresponding frequencies 2*pi/p.

Having the minima and the maxima, the local amplitude can be calculated as the difference between two neighboring minimum and maximum points:

```
def amplitudes(minima, maxima):
    a = [(abs(maxima[n][1] - minima[n][1]))/2.0
         for n in range(min(len(minima),len(maxima)))]
    return np.array(a)
```

The code segments are found in the file vib_empirical_analysis.py.

Since a[i] and p[i] correspond to the i-th amplitude estimate and the i-th period estimate, respectively, it is most convenient to visualize the a and p values with the index i on the horizontal axis. (There is no unique time point associated with either of these estimate since values at two different time points were used in the computations.)

In the analysis of very long time series, it is advantageous to compute and plot p and a instead of u to get an impression of the development of the oscillations. Let us do this for the scaled problem and $\Delta t = 0.1, 0.05, 0.01$. A ready-made function

```
plot_empirical_freq_and_amplitude(u, t, I, w)
```

computes the empirical amplitudes and periods, and creates a plot where the amplitudes and angular frequencies are visualized together with the exact amplitude I and the exact angular frequency w. We can make a little program for creating the plot:

```
from vib_undamped import solver, plot_empirical_freq_and_amplitude
from math import pi
dt_values = [0.1, 0.05, 0.01]
u_cases = []
t_cases = []
for dt in dt_values:
    # Simulate scaled problem for 40 periods
    u, t = solver(I=1, w=2*pi, dt=dt, T=40)
    u_cases.append(u)
    t_cases.append(t)
plot_empirical_freq_and_amplitude(u_cases, t_cases, I=1, w=2*pi)
```

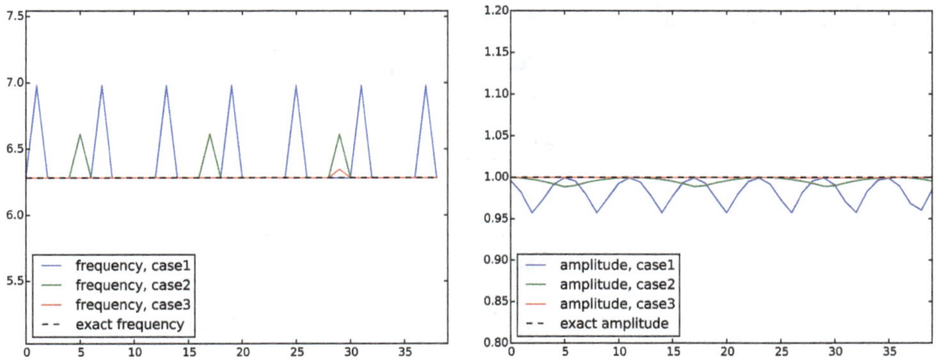

Fig. 1.3 Empirical angular frequency (*left*) and amplitude (*right*) for three different time steps

Figure 1.3 shows the result: we clearly see that lowering Δt improves the angular frequency significantly, while the amplitude seems to be more accurate. The lines with $\Delta t = 0.01$, corresponding to 100 steps per period, can hardly be distinguished from the exact values. The next section shows how we can get mathematical insight into why amplitudes are good while frequencies are more inaccurate.

1.4 Analysis of the Numerical Scheme

1.4.1 Deriving a Solution of the Numerical Scheme

After having seen the phase error grow with time in the previous section, we shall now quantify this error through mathematical analysis. The key tool in the analysis will be to establish an exact solution of the discrete equations. The difference equation (1.7) has constant coefficients and is homogeneous. Such equations are known to have solutions on the form $u^n = CA^n$, where A is some number to be determined from the difference equation and C is found as the initial condition ($C = I$). Recall that n in u^n is a superscript labeling the time level, while n in A^n is an exponent.

With oscillating functions as solutions, the algebra will be considerably simplified if we seek an A on the form

$$A = e^{i\tilde{\omega}\Delta t},$$

and solve for the numerical frequency $\tilde{\omega}$ rather than A. Note that $i = \sqrt{-1}$ is the imaginary unit. (Using a complex exponential function gives simpler arithmetics than working with a sine or cosine function.) We have

$$A^n = e^{i\tilde{\omega}\Delta t\, n} = e^{i\tilde{\omega}t_n} = \cos(\tilde{\omega}t_n) + i\,\sin(\tilde{\omega}t_n)\,.$$

The physically relevant numerical solution can be taken as the real part of this complex expression.

The calculations go as

$$
\begin{aligned}
[D_tD_tu]^n &= \frac{u^{n+1} - 2u^n + u^{n-1}}{\Delta t^2} \\
&= I\frac{A^{n+1} - 2A^n + A^{n-1}}{\Delta t^2} \\
&= \frac{I}{\Delta t^2}\left(e^{i\tilde{\omega}(t_n+\Delta t)} - 2e^{i\tilde{\omega}t_n} + e^{i\tilde{\omega}(t_n-\Delta t)}\right) \\
&= Ie^{i\tilde{\omega}t_n}\frac{1}{\Delta t^2}\left(e^{i\tilde{\omega}\Delta t} + e^{i\tilde{\omega}(-\Delta t)} - 2\right) \\
&= Ie^{i\tilde{\omega}t_n}\frac{2}{\Delta t^2}\left(\cosh(i\tilde{\omega}\Delta t) - 1\right) \\
&= Ie^{i\tilde{\omega}t_n}\frac{2}{\Delta t^2}\left(\cos(\tilde{\omega}\Delta t) - 1\right) \\
&= -Ie^{i\tilde{\omega}t_n}\frac{4}{\Delta t^2}\sin^2\left(\frac{\tilde{\omega}\Delta t}{2}\right).
\end{aligned}
$$

The last line follows from the relation $\cos x - 1 = -2\sin^2(x/2)$ (try `cos(x)-1` in wolframalpha.com[9] to see the formula).

The scheme (1.7) with $u^n = Ie^{i\tilde{\omega}\Delta t\, n}$ inserted now gives

$$
-Ie^{i\tilde{\omega}t_n}\frac{4}{\Delta t^2}\sin^2\left(\frac{\tilde{\omega}\Delta t}{2}\right) + \omega^2 Ie^{i\tilde{\omega}t_n} = 0, \tag{1.16}
$$

which after dividing by $Ie^{i\tilde{\omega}t_n}$ results in

$$
\frac{4}{\Delta t^2}\sin^2\left(\frac{\tilde{\omega}\Delta t}{2}\right) = \omega^2. \tag{1.17}
$$

The first step in solving for the unknown $\tilde{\omega}$ is

$$
\sin^2\left(\frac{\tilde{\omega}\Delta t}{2}\right) = \left(\frac{\omega\Delta t}{2}\right)^2.
$$

Then, taking the square root, applying the inverse sine function, and multiplying by $2/\Delta t$, results in

$$
\tilde{\omega} = \pm\frac{2}{\Delta t}\sin^{-1}\left(\frac{\omega\Delta t}{2}\right). \tag{1.18}
$$

1.4.2 The Error in the Numerical Frequency

The first observation following (1.18) tells that there is a phase error since the numerical frequency $\tilde{\omega}$ never equals the exact frequency ω. But how good is the approximation (1.18)? That is, what is the error $\omega - \tilde{\omega}$ or $\tilde{\omega}/\omega$? Taylor series expansion for small Δt may give an expression that is easier to understand than the complicated function in (1.18):

[9] http://www.wolframalpha.com

```
>>> from sympy import *
>>> dt, w = symbols('dt w')
>>> w_tilde_e = 2/dt*asin(w*dt/2)
>>> w_tilde_series = w_tilde_e.series(dt, 0, 4)
>>> print w_tilde_series
w + dt**2*w**3/24 + O(dt**4)
```

This means that

$$\tilde{\omega} = \omega\left(1 + \frac{1}{24}\omega^2\Delta t^2\right) + \mathcal{O}(\Delta t^4)\,. \tag{1.19}$$

The error in the numerical frequency is of second-order in Δt, and the error vanishes as $\Delta t \to 0$. We see that $\tilde{\omega} > \omega$ since the term $\omega^3\Delta t^2/24 > 0$ and this is by far the biggest term in the series expansion for small $\omega\Delta t$. A numerical frequency that is too large gives an oscillating curve that oscillates too fast and therefore "lags behind" the exact oscillations, a feature that can be seen in the left plot in Fig. 1.2.

Figure 1.4 plots the discrete frequency (1.18) and its approximation (1.19) for $\omega = 1$ (based on the program `vib_plot_freq.py`). Although $\tilde{\omega}$ is a function of Δt in (1.19), it is misleading to think of Δt as the important discretization parameter. It is the product $\omega\Delta t$ that is the key discretization parameter. This quantity reflects the *number of time steps per period* of the oscillations. To see this, we set $P = N_P\Delta t$, where P is the length of a period, and N_P is the number of time steps during a period. Since P and ω are related by $P = 2\pi/\omega$, we get that $\omega\Delta t = 2\pi/N_P$, which shows that $\omega\Delta t$ is directly related to N_P.

The plot shows that at least $N_P \sim 25 - 30$ points per period are necessary for reasonable accuracy, but this depends on the length of the simulation (T) as the total phase error due to the frequency error grows linearly with time (see Exercise 1.2).

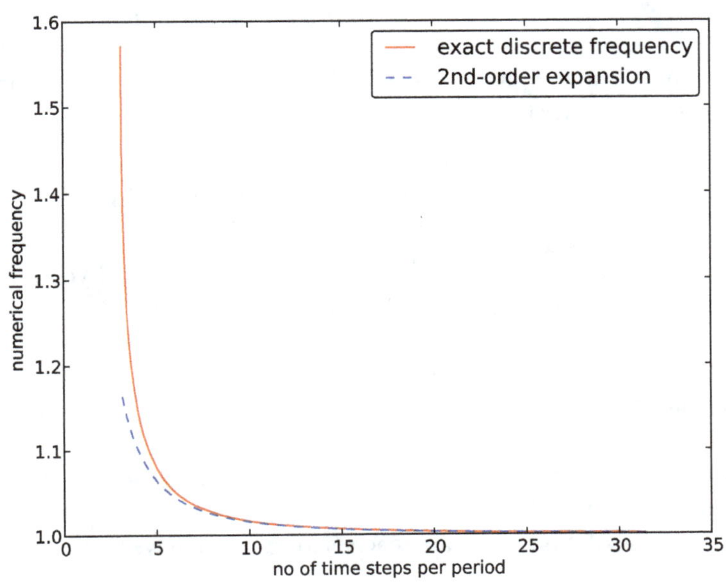

Fig. 1.4 Exact discrete frequency and its second-order series expansion

1.4.3 Empirical Convergence Rates and Adjusted ω

The expression (1.19) suggests that adjusting omega to

$$\omega\left(1-\frac{1}{24}\omega^2\Delta t^2\right),$$

could have effect on the *convergence rate* of the global error in u (cf. Sect. 1.2.2). With the `convergence_rates` function in `vib_undamped.py` we can easily check this. A special solver, with adjusted w, is available as the function `solver_adjust_w`. A call to `convergence_rates` with this solver reveals that the rate is 4.0! With the original, physical ω the rate is 2.0 – as expected from using second-order finite difference approximations, as expected from the forthcoming derivation of the global error, and as expected from truncation error analysis analysis as explained in Appendix B.4.1.

Adjusting ω is an ideal trick for this simple problem, but when adding damping and nonlinear terms, we have no simple formula for the impact on ω, and therefore we cannot use the trick.

1.4.4 Exact Discrete Solution

Perhaps more important than the $\tilde{\omega} = \omega + \mathcal{O}(\Delta t^2)$ result found above is the fact that we have an exact discrete solution of the problem:

$$u^n = I\cos\left(\tilde{\omega}n\Delta t\right), \quad \tilde{\omega} = \frac{2}{\Delta t}\sin^{-1}\left(\frac{\omega\Delta t}{2}\right). \tag{1.20}$$

We can then compute the error mesh function

$$e^n = u_{\mathrm{e}}(t_n) - u^n = I\cos\left(\omega n\Delta t\right) - I\cos\left(\tilde{\omega}n\Delta t\right). \tag{1.21}$$

From the formula $\cos 2x - \cos 2y = -2\sin(x-y)\sin(x+y)$ we can rewrite e^n so the expression is easier to interpret:

$$e^n = -2I\sin\left(t\frac{1}{2}\left(\omega-\tilde{\omega}\right)\right)\sin\left(t\frac{1}{2}\left(\omega+\tilde{\omega}\right)\right). \tag{1.22}$$

The error mesh function is ideal for verification purposes and you are strongly encouraged to make a test based on (1.20) by doing Exercise 1.11.

1.4.5 Convergence

We can use (1.19) and (1.21), or (1.22), to show *convergence* of the numerical scheme, i.e., $e^n \to 0$ as $\Delta t \to 0$, which implies that the numerical solution approaches the exact solution as Δt approaches to zero. We have that

$$\lim_{\Delta t\to 0}\tilde{\omega} = \lim_{\Delta t\to 0}\frac{2}{\Delta t}\sin^{-1}\left(\frac{\omega\Delta t}{2}\right) = \omega,$$

by L'Hopital's rule. This result could also been computed WolframAlpha[10], or we could use the limit functionality in sympy:

```
>>> import sympy as sym
>>> dt, w = sym.symbols('x w')
>>> sym.limit((2/dt)*sym.asin(w*dt/2), dt, 0, dir='+')
w
```

Also (1.19) can be used to establish that $\tilde{\omega} \rightarrow \omega$ when $\Delta t \rightarrow 0$. It then follows from the expression(s) for e^n that $e^n \rightarrow 0$.

1.4.6 The Global Error

To achieve more analytical insight into the nature of the global error, we can Taylor expand the error mesh function (1.21). Since $\tilde{\omega}$ in (1.18) contains Δt in the denominator we use the series expansion for $\tilde{\omega}$ inside the cosine function. A relevant sympy session is

```
>>> from sympy import *
>>> dt, w, t = symbols('dt w t')
>>> w_tilde_e = 2/dt*asin(w*dt/2)
>>> w_tilde_series = w_tilde_e.series(dt, 0, 4)
>>> w_tilde_series
w + dt**2*w**3/24 + O(dt**4)
```

Series expansions in sympy have the inconvenient `O()` term that prevents further calculations with the series. We can use the `removeO()` command to get rid of the `O()` term:

```
>>> w_tilde_series = w_tilde_series.removeO()
>>> w_tilde_series
dt**2*w**3/24 + w
```

Using this `w_tilde_series` expression for $\tilde{w}$ in (1.21), dropping I (which is a common factor), and performing a series expansion of the error yields

```
>>> error = cos(w*t) - cos(w_tilde_series*t)
>>> error.series(dt, 0, 6)
dt**2*t*w**3*sin(t*w)/24 + dt**4*t**2*w**6*cos(t*w)/1152 + O(dt**6)
```

Since we are mainly interested in the leading-order term in such expansions (the term with lowest power in Δt, which goes most slowly to zero), we use the `.as_leading_term(dt)` construction to pick out this term:

[10] http://www.wolframalpha.com/input/?i=%282%2Fx%29*asin%28w*x%2F2%29+as+x-%3E0

```
>>> error.series(dt, 0, 6).as_leading_term(dt)
dt**2*t*w**3*sin(t*w)/24
```

The last result means that the leading order global (true) error at a point t is proportional to $\omega^3 t\Delta t^2$. Considering only the discrete t_n values for t, t_n is related to Δt through $t_n = n\Delta t$. The factor $\sin(\omega t)$ can at most be 1, so we use this value to bound the leading-order expression to its maximum value

$$e^n = \frac{1}{24}n\omega^3\Delta t^3\,.$$

This is the dominating term of the error *at a point*.

We are interested in the accumulated global error, which can be taken as the ℓ^2 norm of e^n. The norm is simply computed by summing contributions from all mesh points:

$$||e^n||_{\ell^2}^2 = \Delta t\sum_{n=0}^{N_t}\frac{1}{24^2}n^2\omega^6\Delta t^6 = \frac{1}{24^2}\omega^6\Delta t^7\sum_{n=0}^{N_t}n^2\,.$$

The sum $\sum_{n=0}^{N_t}n^2$ is approximately equal to $\frac{1}{3}N_t^3$. Replacing N_t by $T/\Delta t$ and taking the square root gives the expression

$$||e^n||_{\ell^2} = \frac{1}{24}\sqrt{\frac{T^3}{3}}\omega^3\Delta t^2\,.$$

This is our expression for the global (or integrated) error. A primary result from this expression is that the global error is proportional to Δt^2.

1.4.7 Stability

Looking at (1.20), it appears that the numerical solution has constant and correct amplitude, but an error in the angular frequency. A constant amplitude is not necessarily the case, however! To see this, note that if only Δt is large enough, the magnitude of the argument to $\sin^{-1}$ in (1.18) may be larger than 1, i.e., $\omega\Delta t/2 > 1$. In this case, $\sin^{-1}(\omega\Delta t/2)$ has a complex value and therefore $\tilde{\omega}$ becomes complex. Type, for example, `asin(x)` in wolframalpha.com[11] to see basic properties of $\sin^{-1}(x)$.

A complex $\tilde{\omega}$ can be written $\tilde{\omega} = \tilde{\omega}_r + i\tilde{\omega}_i$. Since $\sin^{-1}(x)$ has a *negative* imaginary part for $x > 1$, $\tilde{\omega}_i < 0$, which means that $e^{i\tilde{\omega}t} = e^{-\tilde{\omega}_i t}e^{i\tilde{\omega}_r t}$ will lead to exponential growth in time because $e^{-\tilde{\omega}_i t}$ with $\tilde{\omega}_i < 0$ has a positive exponent.

Stability criterion

We do not tolerate growth in the amplitude since such growth is not present in the exact solution. Therefore, we must impose a *stability criterion* so that the

[11] http://www.wolframalpha.com

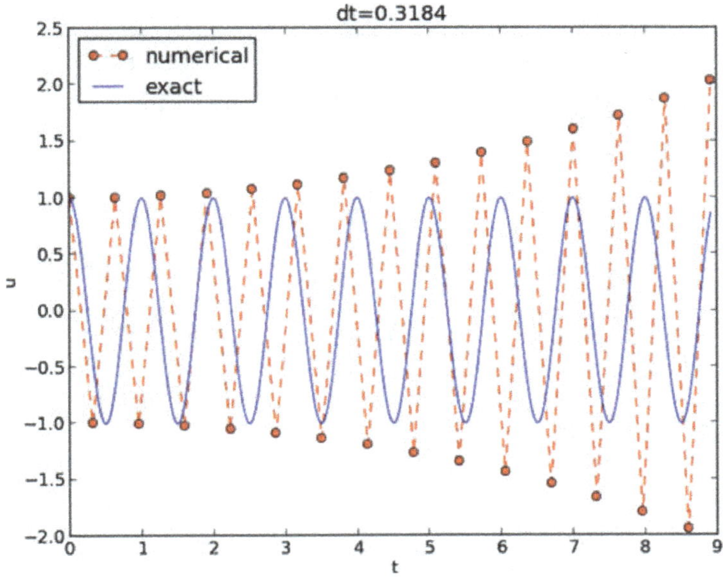

Fig. 1.5 Growing, unstable solution because of a time step slightly beyond the stability limit

argument in the inverse sine function leads to real and not complex values of $\tilde{\omega}$. The stability criterion reads

$$\frac{\omega \Delta t}{2} \leq 1 \quad \Rightarrow \quad \Delta t \leq \frac{2}{\omega}. \tag{1.23}$$

With $\omega = 2\pi$, $\Delta t > \pi^{-1} = 0.3183098861837907$ will give growing solutions. Figure 1.5 displays what happens when $\Delta t = 0.3184$, which is slightly above the critical value: $\Delta t = \pi^{-1} + 9.01 \cdot 10^{-5}$.

1.4.8 About the Accuracy at the Stability Limit

An interesting question is whether the stability condition $\Delta t < 2/\omega$ is unfortunate, or more precisely: would it be meaningful to take larger time steps to speed up computations? The answer is a clear no. At the stability limit, we have that $\sin^{-1} \omega\Delta t/2 = \sin^{-1} 1 = \pi/2$, and therefore $\tilde{\omega} = \pi/\Delta t$. (Note that the approximate formula (1.19) is very inaccurate for this value of Δt as it predicts $\tilde{\omega} = 2.34/pi$, which is a 25 percent reduction.) The corresponding period of the numerical solution is $\tilde{P} = 2\pi/\tilde{\omega} = 2\Delta t$, which means that there is just one time step Δt between a peak (maximum) and a through[12] (minimum) in the numerical solution. This is the shortest possible wave that can be represented in the mesh! In other words, it is not meaningful to use a larger time step than the stability limit.

Also, the error in angular frequency when $\Delta t = 2/\omega$ is severe: Figure 1.6 shows a comparison of the numerical and analytical solution with $\omega = 2\pi$ and

[12] https://simple.wikipedia.org/wiki/Wave_(physics)

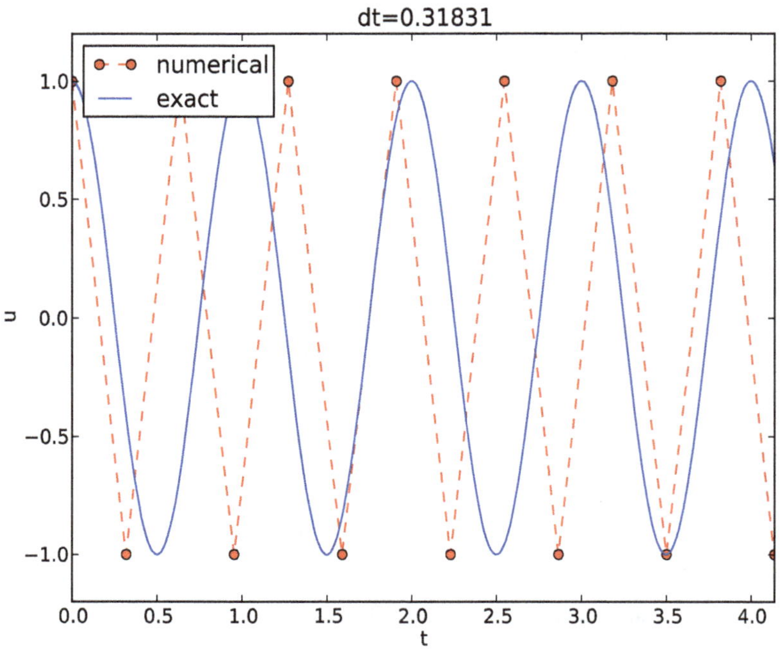

Fig. 1.6 Numerical solution with Δt exactly at the stability limit

$\Delta t = 2/\omega = \pi^{-1}$. Already after one period, the numerical solution has a through while the exact solution has a peak (!). The error in frequency when Δt is at the stability limit becomes $\omega - \tilde{\omega} = \omega(1 - \pi/2) \approx -0.57\omega$. The corresponding error in the period is $P - \tilde{P} \approx 0.36P$. The error after m periods is then $0.36mP$. This error has reached half a period when $m = 1/(2 \cdot 0.36) \approx 1.38$, which theoretically confirms the observations in Fig. 1.6 that the numerical solution is a through ahead of a peak already after one and a half period. Consequently, Δt should be chosen much less than the stability limit to achieve meaningful numerical computations.

Summary

From the accuracy and stability analysis we can draw three important conclusions:

1. The key parameter in the formulas is $p = \omega\Delta t$. The period of oscillations is $P = 2\pi/\omega$, and the number of time steps per period is $N_P = P/\Delta t$. Therefore, $p = \omega\Delta t = 2\pi/N_P$, showing that the critical parameter is the number of time steps per period. The smallest possible N_P is 2, showing that $p \in (0, \pi]$.
2. Provided $p \leq 2$, the amplitude of the numerical solution is constant.
3. The ratio of the numerical angular frequency and the exact one is $\tilde{\omega}/\omega \approx 1 + \frac{1}{24}p^2$. The error $\frac{1}{24}p^2$ leads to wrongly displaced peaks of the numerical solution, and the error in peak location grows linearly with time (see Exercise 1.2).

1.5 Alternative Schemes Based on 1st-Order Equations

A standard technique for solving second-order ODEs is to rewrite them as a system of first-order ODEs and then choose a solution strategy from the vast collection of methods for first-order ODE systems. Given the second-order ODE problem

$$u'' + \omega^2 u = 0, \quad u(0) = I,\ u'(0) = 0,$$

we introduce the auxiliary variable $v = u'$ and express the ODE problem in terms of first-order derivatives of u and v:

$$u' = v, \tag{1.24}$$

$$v' = -\omega^2 u\,. \tag{1.25}$$

The initial conditions become $u(0) = I$ and $v(0) = 0$.

1.5.1 The Forward Euler Scheme

A Forward Euler approximation to our 2×2 system of ODEs (1.24)–(1.25) becomes

$$[D_t^+ u = v]^n, \tag{1.26}$$

$$[D_t^+ v = -\omega^2 u]^n, \tag{1.27}$$

or written out,

$$u^{n+1} = u^n + \Delta t v^n, \tag{1.28}$$

$$v^{n+1} = v^n - \Delta t \omega^2 u^n\,. \tag{1.29}$$

Let us briefly compare this Forward Euler method with the centered difference scheme for the second-order differential equation. We have from (1.28) and (1.29) applied at levels n and $n-1$ that

$$u^{n+1} = u^n + \Delta t v^n = u^n + \Delta t(v^{n-1} - \Delta t \omega^2 u^{n-1})\,.$$

Since from (1.28)

$$v^{n-1} = \frac{1}{\Delta t}(u^n - u^{n-1}),$$

it follows that

$$u^{n+1} = 2u^n - u^{n-1} - \Delta t^2 \omega^2 u^{n-1},$$

which is very close to the centered difference scheme, but the last term is evaluated at t_{n-1} instead of t_n. Rewriting, so that $\Delta t^2\omega^2 u^{n-1}$ appears alone on the right-hand side, and then dividing by Δt^2, the new left-hand side is an approximation to u'' at t_n, while the right-hand side is sampled at t_{n-1}. All terms should be sampled at the same mesh point, so using $\omega^2 u^{n-1}$ instead of $\omega^2 u^n$ points to a kind of mathematical error in the derivation of the scheme. This error turns out to be rather crucial for the accuracy of the Forward Euler method applied to vibration problems (Sect. 1.5.4 has examples).

The reasoning above does not imply that the Forward Euler scheme is not correct, but more that it is almost equivalent to a second-order accurate scheme for the second-order ODE formulation, and that the error committed has to do with a wrong sampling point.

1.5.2 The Backward Euler Scheme

A Backward Euler approximation to the ODE system is equally easy to write up in the operator notation:

$$[D_t^- u = v]^{n+1}, \tag{1.30}$$

$$[D_t^- v = -\omega u]^{n+1}. \tag{1.31}$$

This becomes a coupled system for u^{n+1} and v^{n+1}:

$$u^{n+1} - \Delta t v^{n+1} = u^n, \tag{1.32}$$

$$v^{n+1} + \Delta t \omega^2 u^{n+1} = v^n. \tag{1.33}$$

We can compare (1.32)–(1.33) with the centered scheme (1.7) for the second-order differential equation. To this end, we eliminate v^{n+1} in (1.32) using (1.33) solved with respect to v^{n+1}. Thereafter, we eliminate v^n using (1.32) solved with respect to v^{n+1} and also replacing $n+1$ by n and n by $n-1$. The resulting equation involving only u^{n+1}, u^n, and u^{n-1} can be ordered as

$$\frac{u^{n+1} - 2u^n + u^{n-1}}{\Delta t^2} = -\omega^2 u^{n+1},$$

which has almost the same form as the centered scheme for the second-order differential equation, but the right-hand side is evaluated at u^{n+1} and not u^n. This inconsistent sampling of terms has a dramatic effect on the numerical solution, as we demonstrate in Sect. 1.5.4.

1.5.3 The Crank-Nicolson Scheme

The Crank-Nicolson scheme takes this form in the operator notation:

$$[D_t u = \overline{v}^t]^{n+\frac{1}{2}}, \tag{1.34}$$

$$[D_t v = -\omega^2 \overline{u}^t]^{n+\frac{1}{2}}. \tag{1.35}$$

Writing the equations out and rearranging terms, shows that this is also a coupled system of two linear equations at each time level:

$$u^{n+1} - \frac{1}{2}\Delta t v^{n+1} = u^n + \frac{1}{2}\Delta t v^n, \tag{1.36}$$

$$v^{n+1} + \frac{1}{2}\Delta t \omega^2 u^{n+1} = v^n - \frac{1}{2}\Delta t \omega^2 u^n. \tag{1.37}$$

We may compare also this scheme to the centered discretization of the second-order ODE. It turns out that the Crank-Nicolson scheme is equivalent to the discretization

$$\frac{u^{n+1} - 2u^n + u^{n-1}}{\Delta t^2} = -\omega^2 \frac{1}{4}(u^{n+1} + 2u^n + u^{n-1}) = -\omega^2 u^n + \mathcal{O}(\Delta t^2) \,. \quad (1.38)$$

That is, the Crank-Nicolson is equivalent to (1.7) for the second-order ODE, apart from an extra term of size Δt^2, but this is an error of the same order as in the finite difference approximation on the left-hand side of the equation anyway. The fact that the Crank-Nicolson scheme is so close to (1.7) makes it a much better method than the Forward or Backward Euler methods for vibration problems, as will be illustrated in Sect. 1.5.4.

Deriving (1.38) is a bit tricky. We start with rewriting the Crank-Nicolson equations as follows

$$u^{n+1} - u^n = \frac{1}{2}\Delta t(v^{n+1} + v^n), \quad (1.39)$$

$$v^{n+1} = v^n - \frac{1}{2}\Delta t\omega^2(u^{n+1} + u^n), \quad (1.40)$$

and add the latter at the previous time level as well:

$$v^n = v^{n-1} - \frac{1}{2}\Delta t\omega^2(u^n + u^{n-1}) \,. \quad (1.41)$$

We can also rewrite (1.39) at the previous time level as

$$v^n + v^{n-1} = \frac{2}{\Delta t}(u^n - u^{n-1}) \,. \quad (1.42)$$

Inserting (1.40) for v^{n+1} in (1.39) and (1.41) for v^n in (1.39) yields after some reordering:

$$u^{n+1} - u^n = \frac{1}{2}\left(-\frac{1}{2}\Delta t\omega^2(u^{n+1} + 2u^n + u^{n-1}) + v^n + v^{n-1}\right) .$$

Now, $v^n + v^{n-1}$ can be eliminated by means of (1.42). The result becomes

$$u^{n+1} - 2u^n + u^{n-1} = -\Delta t^2\omega^2 \frac{1}{4}(u^{n+1} + 2u^n + u^{n-1}) \,. \quad (1.43)$$

It can be shown that

$$\frac{1}{4}(u^{n+1} + 2u^n + u^{n-1}) \approx u^n + \mathcal{O}(\Delta t^2),$$

meaning that (1.43) is an approximation to the centered scheme (1.7) for the second-order ODE where the sampling error in the term $\Delta t^2\omega^2 u^n$ is of the same order as the approximation errors in the finite differences, i.e., $\mathcal{O}(\Delta t^2)$. The Crank-Nicolson scheme written as (1.43) therefore has consistent sampling of all terms at the same time point t_n.

1.5.4 Comparison of Schemes

We can easily compare methods like the ones above (and many more!) with the aid of the Odespy[13] package. Below is a sketch of the code.

```
import odespy
import numpy as np

def f(u, t, w=1):
    # v, u numbering for EulerCromer to work well
    v, u = u  # u is array of length 2 holding our [v, u]
    return [-w**2*u, v]

def run_solvers_and_plot(solvers, timesteps_per_period=20,
                         num_periods=1, I=1, w=2*np.pi):
    P = 2*np.pi/w  # duration of one period
    dt = P/timesteps_per_period
    Nt = num_periods*timesteps_per_period
    T = Nt*dt
    t_mesh = np.linspace(0, T, Nt+1)

    legends = []
    for solver in solvers:
        solver.set(f_kwargs={'w': w})
        solver.set_initial_condition([0, I])
        u, t = solver.solve(t_mesh)
```

There is quite some more code dealing with plots also, and we refer to the source file `vib_undamped_odespy.py` for details. Observe that keyword arguments in `f(u,t,w=1)` can be supplied through a solver parameter `f_kwargs` (dictionary of additional keyword arguments to `f`).

Specification of the Forward Euler, Backward Euler, and Crank-Nicolson schemes is done like this:

```
solvers = [
    odespy.ForwardEuler(f),
    # Implicit methods must use Newton solver to converge
    odespy.BackwardEuler(f, nonlinear_solver='Newton'),
    odespy.CrankNicolson(f, nonlinear_solver='Newton'),
    ]
```

The `vib_undamped_odespy.py` program makes two plots of the computed solutions with the various methods in the `solvers` list: one plot with $u(t)$ versus t, and one *phase plane plot* where v is plotted against u. That is, the phase plane plot is the curve $(u(t), v(t))$ parameterized by t. Analytically, $u = I\cos(\omega t)$ and $v = u' = -\omega I\sin(\omega t)$. The exact curve $(u(t), v(t))$ is therefore an ellipse, which often looks like a circle in a plot if the axes are automatically scaled. The important feature, however, is that the exact curve $(u(t), v(t))$ is closed and repeats itself for every period. Not all numerical schemes are capable of doing that, meaning that the amplitude instead shrinks or grows with time.

[13] https://github.com/hplgit/odespy

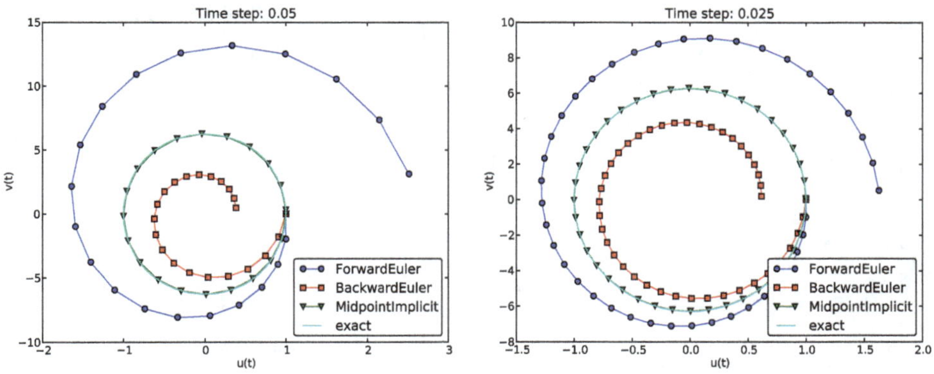

Fig. 1.7 Comparison of classical schemes in the phase plane for two time step values

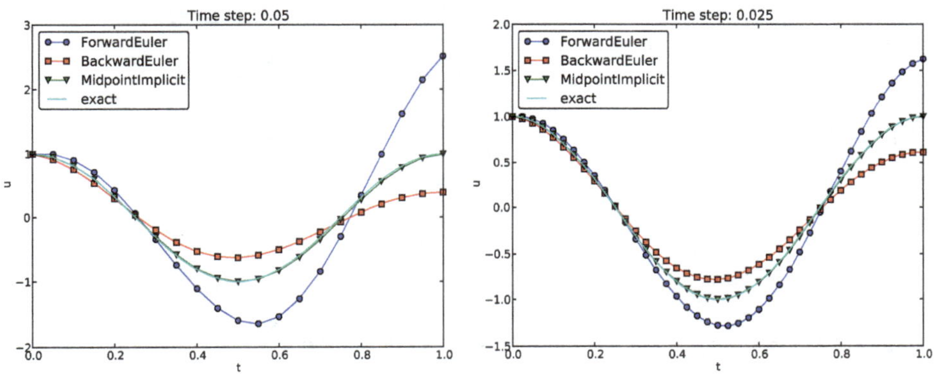

Fig. 1.8 Comparison of solution curves for classical schemes

Figure 1.7 show the results. Note that Odespy applies the label MidpointImplicit for what we have specified as `CrankNicolson` in the code (`CrankNicolson` is just a synonym for class `MidpointImplicit` in the Odespy code). The Forward Euler scheme in Fig. 1.7 has a pronounced spiral curve, pointing to the fact that the amplitude steadily grows, which is also evident in Fig. 1.8. The Backward Euler scheme has a similar feature, except that the spriral goes inward and the amplitude is significantly damped. The changing amplitude and the spiral form decreases with decreasing time step. The Crank-Nicolson scheme looks much more accurate. In fact, these plots tell that the Forward and Backward Euler schemes are not suitable for solving our ODEs with oscillating solutions.

1.5.5 Runge-Kutta Methods

We may run two other popular standard methods for first-order ODEs, the 2nd- and 4th-order Runge-Kutta methods, to see how they perform. Figures 1.9 and 1.10 show the solutions with larger Δt values than what was used in the previous two plots.

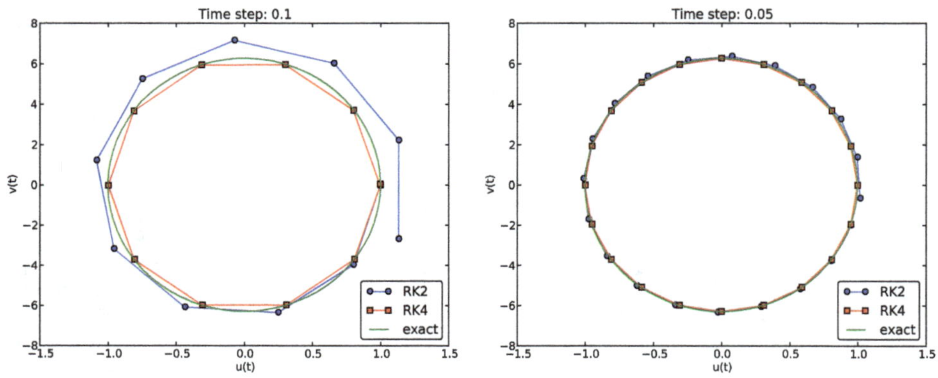

Fig. 1.9 Comparison of Runge-Kutta schemes in the phase plane

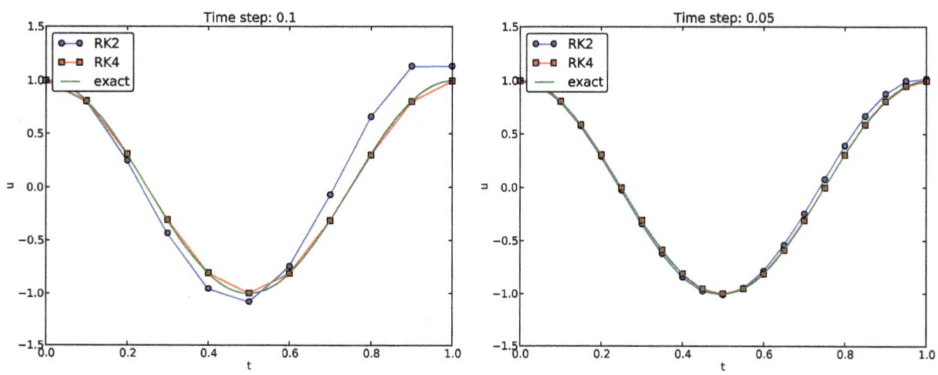

Fig. 1.10 Comparison of Runge-Kutta schemes

The visual impression is that the 4th-order Runge-Kutta method is very accurate, under all circumstances in these tests, while the 2nd-order scheme suffers from amplitude errors unless the time step is very small.

The corresponding results for the Crank-Nicolson scheme are shown in Fig. 1.11. It is clear that the Crank-Nicolson scheme outperforms the 2nd-order Runge-Kutta method. Both schemes have the same order of accuracy $\mathcal{O}(\Delta t^2)$, but their differences in the accuracy that matters in a real physical application is very clearly pronounced in this example. Exercise 1.13 invites you to investigate how the amplitude is computed by a series of famous methods for first-order ODEs.

1.5.6 Analysis of the Forward Euler Scheme

We may try to find exact solutions of the discrete equations (1.28)–(1.29) in the Forward Euler method to better understand why this otherwise useful method has so bad performance for vibration ODEs. An "ansatz" for the solution of the discrete equations is

$$u^n = IA^n,$$
$$v^n = qIA^n,$$

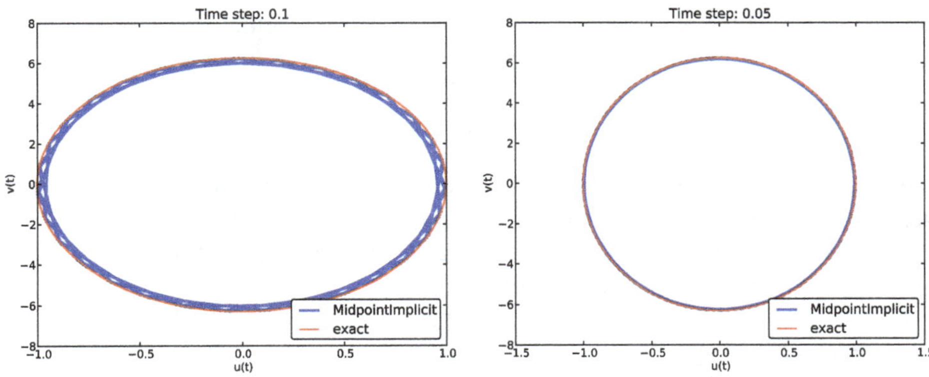

Fig. 1.11 Long-time behavior of the Crank-Nicolson scheme in the phase plane

where q and A are scalars to be determined. We could have used a complex exponential form $e^{i\tilde{\omega}n\Delta t}$ since we get oscillatory solutions, but the oscillations grow in the Forward Euler method, so the numerical frequency $\tilde{\omega}$ will be complex anyway (producing an exponentially growing amplitude). Therefore, it is easier to just work with potentially complex A and q as introduced above.

The Forward Euler scheme leads to

$$
\begin{aligned}
A &= 1 + \Delta t q, \\
A &= 1 - \Delta t \omega^2 q^{-1}.
\end{aligned}
$$

We can easily eliminate A, get $q^2 + \omega^2 = 0$, and solve for

$$q = \pm i\omega,$$

which gives

$$A = 1 \pm \Delta t i\omega.$$

We shall take the real part of A^n as the solution. The two values of A are complex conjugates, and the real part of A^n will be the same for both roots. This is easy to realize if we rewrite the complex numbers in polar form, which is also convenient for further analysis and understanding. The polar form $re^{i\theta}$ of a complex number $x + iy$ has $r = \sqrt{x^2 + y^2}$ and $\theta = \tan^{-1}(y/x)$. Hence, the polar form of the two values for A becomes

$$1 \pm \Delta t i\omega = \sqrt{1 + \omega^2 \Delta t^2} e^{\pm i \tan^{-1}(\omega \Delta t)}.$$

Now it is very easy to compute A^n:

$$(1 \pm \Delta t i\omega)^n = (1 + \omega^2 \Delta t^2)^{n/2} e^{\pm n i \tan^{-1}(\omega \Delta t)}.$$

Since $\cos(\theta n) = \cos(-\theta n)$, the real parts of the two numbers become the same. We therefore continue with the solution that has the plus sign.

The general solution is $u^n = CA^n$, where C is a constant determined from the initial condition: $u^0 = C = I$. We have $u^n = IA^n$ and $v^n = qIA^n$. The final solutions are just the real part of the expressions in polar form:

$$u^n = I(1+\omega^2\Delta t^2)^{n/2}\cos(n\tan^{-1}(\omega\Delta t)), \tag{1.44}$$

$$v^n = -\omega I(1+\omega^2\Delta t^2)^{n/2}\sin(n\tan^{-1}(\omega\Delta t))\,. \tag{1.45}$$

The expression $(1+\omega^2\Delta t^2)^{n/2}$ causes growth of the amplitude, since a number greater than one is raised to a positive exponent $n/2$. We can develop a series expression to better understand the formula for the amplitude. Introducing $p = \omega\Delta t$ as the key variable and using sympy gives

```
>>> from sympy import *
>>> p = symbols('p', real=True)
>>> n = symbols('n', integer=True, positive=True)
>>> amplitude = (1 + p**2)**(n/2)
>>> amplitude.series(p, 0, 4)
1 + n*p**2/2 + O(p**4)
```

The amplitude goes like $1 + \frac{1}{2}n\omega^2\Delta t^2$, clearly growing linearly in time (with n).

We can also investigate the error in the angular frequency by a series expansion:

```
>>> n*atan(p).series(p, 0, 4)
n*(p - p**3/3 + O(p**4))
```

This means that the solution for u^n can be written as

$$u^n = \left(1 + \frac{1}{2}n\omega^2\Delta t^2 + \mathcal{O}(\Delta t^4)\right)\cos\left(\omega t - \frac{1}{3}\omega t\Delta t^2 + \mathcal{O}(\Delta t^4)\right).$$

The error in the angular frequency is of the same order as in the scheme (1.7) for the second-order ODE, but the error in the amplitude is severe.

1.6 Energy Considerations

The observations of various methods in the previous section can be better interpreted if we compute a quantity reflecting the total *energy of the system*. It turns out that this quantity,

$$E(t) = \frac{1}{2}(u')^2 + \frac{1}{2}\omega^2 u^2,$$

is *constant* for all t. Checking that $E(t)$ really remains constant brings evidence that the numerical computations are sound. It turns out that E is proportional to the mechanical energy in the system. Conservation of energy is much used to check numerical simulations, so it is well invested time to dive into this subject.

1.6.1 Derivation of the Energy Expression

We start out with multiplying

$$u'' + \omega^2 u = 0,$$

by u' and integrating from 0 to T:

$$\int_0^T u''u'dt + \int_0^T \omega^2 uu'dt = 0\,.$$

Observing that

$$u''u' = \frac{d}{dt}\frac{1}{2}(u')^2, \quad uu' = \frac{d}{dt}\frac{1}{2}u^2,$$

we get

$$\int_0^T \left(\frac{d}{dt}\frac{1}{2}(u')^2 + \frac{d}{dt}\frac{1}{2}\omega^2 u^2\right) dt = E(T) - E(0) = 0,$$

where we have introduced

$$E(t) = \frac{1}{2}(u')^2 + \frac{1}{2}\omega^2 u^2\,. \tag{1.46}$$

The important result from this derivation is that the total energy is constant:

$$E(t) = E(0)\,.$$

$E(t)$ is closely related to the system's energy

The quantity $E(t)$ derived above is physically not the mechanical energy of a vibrating mechanical system, but the energy per unit mass. To see this, we start with Newton's second law $F = ma$ (F is the sum of forces, m is the mass of the system, and a is the acceleration). The displacement u is related to a through $a = u''$. With a spring force as the only force we have $F = -ku$, where k is a spring constant measuring the stiffness of the spring. Newton's second law then implies the differential equation

$$-ku = mu'' \quad \Rightarrow \quad mu'' + ku = 0\,.$$

This equation of motion can be turned into an energy balance equation by finding the work done by each term during a time interval $[0, T]$. To this end, we multiply the equation by $du = u'dt$ and integrate:

$$\int_0^T muu'dt + \int_0^T kuu'dt = 0\,.$$

The result is

$$\tilde{E}(t) = E_k(t) + E_p(t) = 0,$$

where

$$E_k(t) = \frac{1}{2}mv^2, \quad v = u', \tag{1.47}$$

is the *kinetic energy* of the system, and

$$E_p(t) = \frac{1}{2}ku^2 \tag{1.48}$$

is the *potential energy*. The sum $\tilde{E}(t)$ is the total mechanical energy. The derivation demonstrates the famous energy principle that, under the right physical circumstances, any change in the kinetic energy is due to a change in potential energy and vice versa. (This principle breaks down when we introduce damping in the system, as we do in Sect. 1.10.)

The equation $mu''+ku = 0$ can be divided by m and written as $u''+\omega^2 u = 0$ for $\omega = \sqrt{k/m}$. The energy expression $E(t) = \frac{1}{2}(u')^2 + \frac{1}{2}\omega^2 u^2$ derived earlier is then $\tilde{E}(t)/m$, i.e., mechanical energy per unit mass.

Energy of the exact solution Analytically, we have $u(t) = I\cos\omega t$, if $u(0) = I$ and $u'(0) = 0$, so we can easily check the energy evolution and confirm that $E(t)$ is constant:

$$E(t) = \frac{1}{2}I^2(-\omega\sin\omega t)^2 + \frac{1}{2}\omega^2 I^2\cos^2\omega t = \frac{1}{2}\omega^2(\sin^2\omega t + \cos^2\omega t) = \frac{1}{2}\omega^2\,.$$

Growth of energy in the Forward Euler scheme It is easy to show that the energy in the Forward Euler scheme increases when stepping from time level n to $n+1$.

$$\begin{aligned}
E^{n+1} &= \frac{1}{2}(v^{n+1})^2 + \frac{1}{2}\omega^2(u^{n+1})^2 \\
&= \frac{1}{2}(v^n - \omega^2\Delta t u^n)^2 + \frac{1}{2}\omega^2(u^n + \Delta t v^n)^2 \\
&= (1 + \Delta t^2\omega^2)E^n\,.
\end{aligned}$$

1.6.2 An Error Measure Based on Energy

The constant energy is well expressed by its initial value $E(0)$, so that the error in mechanical energy can be computed as a mesh function by

$$e_E^n = \frac{1}{2}\left(\frac{u^{n+1} - u^{n-1}}{2\Delta t}\right)^2 + \frac{1}{2}\omega^2(u^n)^2 - E(0), \quad n = 1,\ldots,N_t - 1, \tag{1.49}$$

where

$$E(0) = \frac{1}{2}V^2 + \frac{1}{2}\omega^2 I^2,$$

if $u(0) = I$ and $u'(0) = V$. Note that we have used a centered approximation to u': $u'(t_n) \approx [D_{2t}u]^n$.

A useful norm of the mesh function e_E^n for the discrete mechanical energy can be the maximum absolute value of e_E^n:

$$||e_E^n||_{\ell^\infty} = \max_{1\le n<N_t} |e_E^n|\,.$$

Alternatively, we can compute other norms involving integration over all mesh points, but we are often interested in worst case deviation of the energy, and then the maximum value is of particular relevance.

A vectorized Python implementation of e_E^n takes the form

```
# import numpy as np and compute u, t
dt = t[1]-t[0]
E = 0.5*((u[2:] - u[:-2])/(2*dt))**2 + 0.5*w**2*u[1:-1]**2
E0 = 0.5*V**2 + 0.5**w**2*I**2
e_E = E - E0
e_E_norm = np.abs(e_E).max()
```

The convergence rates of the quantity `e_E_norm` can be used for verification. The value of `e_E_norm` is also useful for comparing schemes through their ability to preserve energy. Below is a table demonstrating the relative error in total energy for various schemes (computed by the `vib_undamped_odespy.py` program). The test problem is $u'' + 4\pi^2 u = 0$ with $u(0) = 1$ and $u'(0) = 0$, so the period is 1 and $E(t) \approx 4.93$. We clearly see that the Crank-Nicolson and the Runge-Kutta schemes are superior to the Forward and Backward Euler schemes already after one period.

Method	T	Δt	$\max \left\lvert e_E^n \right\rvert / e_E^0$
Forward Euler	1	0.025	$1.678 \cdot 10^{0}$
Backward Euler	1	0.025	$6.235 \cdot 10^{-1}$
Crank-Nicolson	1	0.025	$1.221 \cdot 10^{-2}$
Runge-Kutta 2nd-order	1	0.025	$6.076 \cdot 10^{-3}$
Runge-Kutta 4th-order	1	0.025	$8.214 \cdot 10^{-3}$

However, after 10 periods, the picture is much more dramatic:

Method	T	Δt	$\max \left\lvert e_E^n \right\rvert / e_E^0$
Forward Euler	10	0.025	$1.788 \cdot 10^{4}$
Backward Euler	10	0.025	$1.000 \cdot 10^{0}$
Crank-Nicolson	10	0.025	$1.221 \cdot 10^{-2}$
Runge-Kutta 2nd-order	10	0.025	$6.250 \cdot 10^{-2}$
Runge-Kutta 4th-order	10	0.025	$8.288 \cdot 10^{-3}$

The Runge-Kutta and Crank-Nicolson methods hardly change their energy error with T, while the error in the Forward Euler method grows to huge levels and a relative error of 1 in the Backward Euler method points to $E(t) \to 0$ as t grows large.

Running multiple values of Δt, we can get some insight into the convergence of the energy error:

Method	T	Δt	$\max \lvert e_E^n \rvert / e_E^0$
Forward Euler	10	0.05	$1.120 \cdot 10^8$
Forward Euler	10	0.025	$1.788 \cdot 10^4$
Forward Euler	10	0.0125	$1.374 \cdot 10^2$
Backward Euler	10	0.05	$1.000 \cdot 10^0$
Backward Euler	10	0.025	$1.000 \cdot 10^0$
Backward Euler	10	0.0125	$9.928 \cdot 10^{-1}$
Crank-Nicolson	10	0.05	$4.756 \cdot 10^{-2}$
Crank-Nicolson	10	0.025	$1.221 \cdot 10^{-2}$
Crank-Nicolson	10	0.0125	$3.125 \cdot 10^{-3}$
Runge-Kutta 2nd-order	10	0.05	$6.152 \cdot 10^{-1}$
Runge-Kutta 2nd-order	10	0.025	$6.250 \cdot 10^{-2}$
Runge-Kutta 2nd-order	10	0.0125	$7.631 \cdot 10^{-3}$
Runge-Kutta 4th-order	10	0.05	$3.510 \cdot 10^{-2}$
Runge-Kutta 4th-order	10	0.025	$8.288 \cdot 10^{-3}$
Runge-Kutta 4th-order	10	0.0125	$2.058 \cdot 10^{-3}$

A striking fact from this table is that the error of the Forward Euler method is reduced by the same factor as Δt is reduced by, while the error in the Crank-Nicolson method has a reduction proportional to Δt^2 (we cannot say anything for the Backward Euler method). However, for the RK2 method, halving Δt reduces the error by almost a factor of 10 (!), and for the RK4 method the reduction seems proportional to Δt^2 only (and the trend is confirmed by running smaller time steps, so for $\Delta t = 3.9 \cdot 10^{-4}$ the relative error of RK2 is a factor 10 smaller than that of RK4!).

1.7 The Euler-Cromer Method

While the Runge-Kutta methods and the Crank-Nicolson scheme work well for the vibration equation modeled as a first-order ODE system, both were inferior to the straightforward centered difference scheme for the second-order equation $u'' + \omega^2 u = 0$. However, there is a similarly successful scheme available for the first-order system $u' = v$, $v' = -\omega^2 u$, to be presented below. The ideas of the scheme and their further developments have become very popular in particle and rigid body dynamics and hence are widely used by physicists.

1.7.1 Forward-Backward Discretization

The idea is to apply a Forward Euler discretization to the first equation and a Backward Euler discretization to the second. In operator notation this is stated as

$$[D_t^+ u = v]^n, \tag{1.50}$$

$$[D_t^- v = -\omega^2 u]^{n+1}\,. \tag{1.51}$$

We can write out the formulas and collect the unknowns on the left-hand side:

$$u^{n+1} = u^n + \Delta t v^n, \tag{1.52}$$

$$v^{n+1} = v^n - \Delta t \omega^2 u^{n+1}. \tag{1.53}$$

We realize that after u^{n+1} has been computed from (1.52), it may be used directly in (1.53) to compute v^{n+1}.

In physics, it is more common to update the v equation first, with a forward difference, and thereafter the u equation, with a backward difference that applies the most recently computed v value:

$$v^{n+1} = v^n - \Delta t \omega^2 u^n, \tag{1.54}$$

$$u^{n+1} = u^n + \Delta t v^{n+1}. \tag{1.55}$$

The advantage of ordering the ODEs as in (1.54)–(1.55) becomes evident when considering complicated models. Such models are included if we write our vibration ODE more generally as

$$u'' + g(u, u', t) = 0.$$

We can rewrite this second-order ODE as two first-order ODEs,

$$v' = -g(u, v, t),$$
$$u' = v.$$

This rewrite allows the following scheme to be used:

$$v^{n+1} = v^n - \Delta t\, g(u^n, v^n, t),$$
$$u^{n+1} = u^n + \Delta t\, v^{n+1}.$$

We realize that the first update works well with any g since old values u^n and v^n are used. Switching the equations would demand u^{n+1} and v^{n+1} values in g and result in nonlinear algebraic equations to be solved at each time level.

The scheme (1.54)–(1.55) goes under several names: forward-backward scheme, semi-implicit Euler method[14], semi-explicit Euler, symplectic Euler, Newton-Störmer-Verlet, and Euler-Cromer. We shall stick to the latter name.

How does the Euler-Cromer method preserve the total energy? We may run the example from Sect. 1.6.2:

Method	T	Δt	$\max \left\lvert e_E^n \right\rvert / e_E^0$
Euler-Cromer	10	0.05	$2.530 \cdot 10^{-2}$
Euler-Cromer	10	0.025	$6.206 \cdot 10^{-3}$
Euler-Cromer	10	0.0125	$1.544 \cdot 10^{-3}$

The relative error in the total energy decreases as Δt^2, and the error level is slightly lower than for the Crank-Nicolson and Runge-Kutta methods.

[14] http://en.wikipedia.org/wiki/Semi-implicit_Euler_method

1.7.2 Equivalence with the Scheme for the Second-Order ODE

We shall now show that the Euler-Cromer scheme for the system of first-order equations is equivalent to the centered finite difference method for the second-order vibration ODE (!).

We may eliminate the v^n variable from (1.52)–(1.53) or (1.54)–(1.55). The v^{n+1} term in (1.54) can be eliminated from (1.55):

$$u^{n+1} = u^n + \Delta t(v^n - \omega^2\Delta t u^n)\,. \tag{1.56}$$

The v^n quantity can be expressed by u^n and u^{n-1} using (1.55):

$$v^n = \frac{u^n - u^{n-1}}{\Delta t},$$

and when this is inserted in (1.56) we get

$$u^{n+1} = 2u^n - u^{n-1} - \Delta t^2\omega^2 u^n, \tag{1.57}$$

which is nothing but the centered scheme (1.7)! The two seemingly different numerical methods are mathematically equivalent. Consequently, the previous analysis of (1.7) also applies to the Euler-Cromer method. In particular, the amplitude is constant, given that the stability criterion is fulfilled, but there is always an angular frequency error (1.19). Exercise 1.18 gives guidance on how to derive the exact discrete solution of the two equations in the Euler-Cromer method.

Although the Euler-Cromer scheme and the method (1.7) are equivalent, there could be differences in the way they handle the initial conditions. Let us look into this topic. The initial condition $u' = 0$ means $u' = v = 0$. From (1.54) we get

$$v^1 = v^0 - \Delta t\omega^2 u^0 = \Delta t\omega^2 u^0,$$

and from (1.55) it follows that

$$u^1 = u^0 + \Delta t v^1 = u^0 - \omega^2\Delta t^2 u^0\,.$$

When we previously used a centered approximation of $u'(0) = 0$ combined with the discretization (1.7) of the second-order ODE, we got a slightly different result: $u^1 = u^0 - \frac{1}{2}\omega^2\Delta t^2 u^0$. The difference is $\frac{1}{2}\omega^2\Delta t^2 u^0$, which is of second order in Δt, seemingly consistent with the overall error in the scheme for the differential equation model.

A different view can also be taken. If we approximate $u'(0) = 0$ by a backward difference, $(u^0 - u^{-1})/\Delta t = 0$, we get $u^{-1} = u^0$, and when combined with (1.7), it results in $u^1 = u^0 - \omega^2\Delta t^2 u^0$. This means that the Euler-Cromer method based on (1.55)–(1.54) corresponds to using only a first-order approximation to the initial condition in the method from Sect. 1.1.2.

Correspondingly, using the formulation (1.52)–(1.53) with $v^n = 0$ leads to $u^1 = u^0$, which can be interpreted as using a forward difference approximation for the initial condition $u'(0) = 0$. Both Euler-Cromer formulations lead to slightly different values for u^1 compared to the method in Sect. 1.1.2. The error is $\frac{1}{2}\omega^2\Delta t^2 u^0$.

1.7.3 Implementation

Solver function The function below, found in vib_undamped_EulerCromer.py, implements the Euler-Cromer scheme (1.54)–(1.55):

```
import numpy as np

def solver(I, w, dt, T):
    """
    Solve v' = - w**2*u, u'=v for t in (0,T], u(0)=I and v(0)=0,
    by an Euler-Cromer method.
    """
    dt = float(dt)
    Nt = int(round(T/dt))
    u = np.zeros(Nt+1)
    v = np.zeros(Nt+1)
    t = np.linspace(0, Nt*dt, Nt+1)

    v[0] = 0
    u[0] = I
    for n in range(0, Nt):
        v[n+1] = v[n] - dt*w**2*u[n]
        u[n+1] = u[n] + dt*v[n+1]
    return u, v, t
```

Verification Since the Euler-Cromer scheme is equivalent to the finite difference method for the second-order ODE $u'' + \omega^2 u = 0$ (see Sect. 1.7.2), the performance of the above solver function is the same as for the solver function in Sect. 1.2. The only difference is the formula for the first time step, as discussed above. This deviation in the Euler-Cromer scheme means that the discrete solution listed in Sect. 1.4.4 is not a solution of the Euler-Cromer scheme!

To verify the implementation of the Euler-Cromer method we can adjust v[1] so that the computer-generated values can be compared with the formula (1.20) from in Sect. 1.4.4. This adjustment is done in an alternative solver function, solver_ic_fix in vib_EulerCromer.py. Since we now have an exact solution of the discrete equations available, we can write a test function test_solver for checking the equality of computed values with the formula (1.20):

```
def test_solver():
    """
    Test solver with fixed initial condition against
    equivalent scheme for the 2nd-order ODE u'' + u = 0.
    """
    I = 1.2; w = 2.0; T = 5
    dt = 2/w  # longest possible time step
    u, v, t = solver_ic_fix(I, w, dt, T)
    from vib_undamped import solver as solver2  # 2nd-order ODE
    u2, t2 = solver2(I, w, dt, T)
    error = np.abs(u - u2).max()
    tol = 1E-14
    assert error < tol
```

Another function, demo, visualizes the difference between the Euler-Cromer scheme and the scheme (1.7) for the second-oder ODE, arising from the mismatch in the first time level.

Using Odespy The Euler-Cromer method is also available in the Odespy package. The important thing to remember, when using this implementation, is that we must order the unknowns as v and u, so the u vector at each time level consists of the velocity v as first component and the displacement u as second component:

```
# Define ODE
def f(u, t, w=1):
    v, u = u
    return [-w**2*u, v]

# Initialize solver
I = 1
w = 2*np.pi
import odespy
solver = odespy.EulerCromer(f, f_kwargs={'w': w})
solver.set_initial_condition([0, I])

# Compute time mesh
P = 2*np.pi/w  # duration of one period
dt = P/timesteps_per_period
Nt = num_periods*timesteps_per_period
T = Nt*dt
import numpy as np
t_mesh = np.linspace(0, T, Nt+1)

# Solve ODE
u, t = solver.solve(t_mesh)
u = u[:,1]  # Extract displacement
```

Convergence rates We may use the convergence_rates function in the file vib_undamped.py to investigate the convergence rate of the Euler-Cromer method, see the convergence_rate function in the file vib_undamped_EulerCromer.py. Since we could eliminate v to get a scheme for u that is equivalent to the finite difference method for the second-order equation in u, we would expect the convergence rates to be the same, i.e., $r = 2$. However, measuring the convergence rate of u in the Euler-Cromer scheme shows that $r = 1$ only! Adjusting the initial condition does not change the rate. Adjusting ω, as outlined in Sect. 1.4.2, gives a 4th-order method there, while there is no increase in the measured rate in the Euler-Cromer scheme. It is obvious that the Euler-Cromer scheme is dramatically much better than the two other first-order methods, Forward Euler and Backward Euler, but this is not reflected in the convergence rate of u.

1.7.4 The Störmer-Verlet Algorithm

Another very popular algorithm for vibration problems, especially for long time simulations, is the Störmer-Verlet algorithm. It has become *the* method among physicists for molecular simulations as well as particle and rigid body dynamics.

The method can be derived by applying the Euler-Cromer idea twice, in a symmetric fashion, during the interval $[t_n, t_{n+1}]$:

1. solve $v' = -\omega u$ by a Forward Euler step in $[t_n, t_{n+\frac{1}{2}}]$
2. solve $u' = v$ by a Backward Euler step in $[t_n, t_{n+\frac{1}{2}}]$
3. solve $u' = v$ by a Forward Euler step in $[t_{n+\frac{1}{2}}, t_{n+1}]$
4. solve $v' = -\omega u$ by a Backward Euler step in $[t_{n+\frac{1}{2}}, t_{n+1}]$

With mathematics,

$$\begin{aligned}
\frac{v^{n+\frac{1}{2}} - v^n}{\frac{1}{2}\Delta t} &= -\omega^2 u^n, \\
\frac{u^{n+\frac{1}{2}} - u^n}{\frac{1}{2}\Delta t} &= v^{n+\frac{1}{2}}, \\
\frac{u^{n+1} - u^{n+\frac{1}{2}}}{\frac{1}{2}\Delta t} &= v^{n+\frac{1}{2}}, \\
\frac{v^{n+1} - v^{n+\frac{1}{2}}}{\frac{1}{2}\Delta t} &= -\omega^2 u^{n+1}.
\end{aligned}$$

The two steps in the middle can be combined to

$$\frac{u^{n+1} - u^n}{\Delta t} = v^{n+\frac{1}{2}},$$

and consequently

$$v^{n+\frac{1}{2}} = v^n - \frac{1}{2}\Delta t\omega^2 u^n, \tag{1.58}$$

$$u^{n+1} = u^n + \Delta t v^{n+\frac{1}{2}}, \tag{1.59}$$

$$v^{n+1} = v^{n+\frac{1}{2}} - \frac{1}{2}\Delta t\omega^2 u^{n+1}. \tag{1.60}$$

Writing the last equation as $v^n = v^{n-\frac{1}{2}} - \frac{1}{2}\Delta t\omega^2 u^n$ and using this v^n in the first equation gives $v^{n+\frac{1}{2}} = v^{n-\frac{1}{2}} - \Delta t\omega^2 u^n$, and the scheme can be written as two steps:

$$v^{n+\frac{1}{2}} = v^{n-\frac{1}{2}} - \Delta t\omega^2 u^n, \tag{1.61}$$

$$u^{n+1} = u^n + \Delta t v^{n+\frac{1}{2}}, \tag{1.62}$$

which is nothing but straightforward centered differences for the 2×2 ODE system on a *staggered mesh*, see Sect. 1.8.1. We have thus seen that four different reasonings (discretizing $u'' + \omega^2 u$ directly, using Euler-Cromer, using Stömer-Verlet, and using centered differences for the 2×2 system on a staggered mesh) all end up with the same equations! The main difference is that the traditional Euler-Cromer displays first-order convergence in Δt (due to less symmetry in the way u and v are treated) while the others are $\mathcal{O}(\Delta t^2)$ schemes.

The most numerically stable scheme, with respect to accumulation of rounding errors, is (1.61)–(1.62). It has, according to [6], better properties in this regard than the direct scheme for the second-order ODE.

1.8 Staggered Mesh

A more intuitive discretization than the Euler-Cromer method, yet equivalent, employs solely centered differences in a natural way for the 2×2 first-order ODE system. The scheme is in fact fully equivalent to the second-order scheme for $u'' + \omega u = 0$, also for the first time step. Such a scheme needs to operate on a *staggered mesh* in time. Staggered meshes are very popular in many physical application, maybe foremost fluid dynamics and electromagnetics, so the topic is important to learn.

1.8.1 The Euler-Cromer Scheme on a Staggered Mesh

In a staggered mesh, the unknowns are sought at different points in the mesh. Specifically, u is sought at integer time points t_n and v is sought at $t_{n+1/2}$ *between* two u points. The unknowns are then $u^1, v^{3/2}, u^2, v^{5/2}$, and so on. We typically use the notation u^n and $v^{n+\frac{1}{2}}$ for the two unknown mesh functions. Figure 1.12 presents a graphical sketch of two mesh functions u and v on a staggered mesh.

On a staggered mesh it is natural to use centered difference approximations, expressed in operator notation as

$$[D_t u = v]^{n+\frac{1}{2}}, \tag{1.63}$$

$$[D_t v = -\omega^2 u]^{n+1}, \tag{1.64}$$

or if we switch the sequence of the equations:

$$[D_t v = -\omega^2 u]^{n}, \tag{1.65}$$

$$[D_t u = v]^{n+\frac{1}{2}}\,. \tag{1.66}$$

Writing out the formulas gives

$$v^{n+\frac{1}{2}} = v^{n-\frac{1}{2}} - \Delta t \omega^2 u^n, \tag{1.67}$$

$$u^{n+1} = u^n + \Delta t v^{n+\frac{1}{2}}\,. \tag{1.68}$$

We can eliminate the v values and get back the centered scheme based on the second-order differential equation $u'' + \omega^2 u = 0$, so all these three schemes are equivalent. However, they differ somewhat in the treatment of the initial conditions.

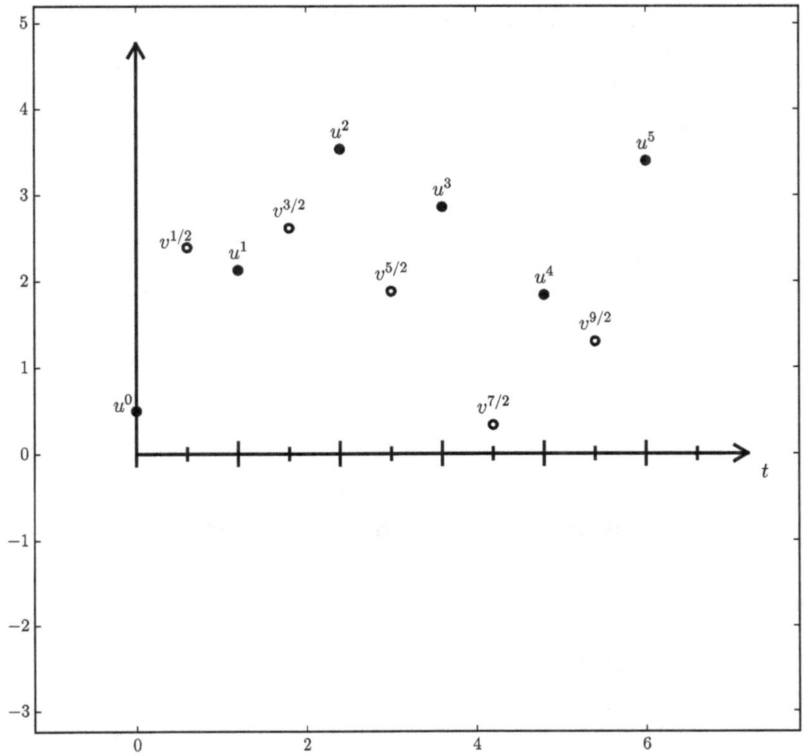

Fig. 1.12 Examples on mesh functions on a staggered mesh in time

Suppose we have $u(0) = I$ and $u'(0) = v(0) = 0$ as mathematical initial conditions. This means $u^0 = I$ and

$$v(0) \approx \frac{1}{2}\left(v^{-\frac{1}{2}} + v^{\frac{1}{2}}\right) = 0, \quad \Rightarrow \quad v^{-\frac{1}{2}} = -v^{\frac{1}{2}}\,.$$

Using the discretized equation (1.67) for $n = 0$ yields

$$v^{\frac{1}{2}} = v^{-\frac{1}{2}} - \Delta t\omega^2 I,$$

and eliminating $v^{-\frac{1}{2}} = -v^{\frac{1}{2}}$ results in

$$v^{\frac{1}{2}} = -\frac{1}{2}\Delta t\omega^2 I,$$

and

$$u^1 = u^0 - \frac{1}{2}\Delta t^2\omega^2 I,$$

which is exactly the same equation for u^1 as we had in the centered scheme based on the second-order differential equation (and hence corresponds to a centered difference approximation of the initial condition for $u'(0)$). The conclusion is that a staggered mesh is fully equivalent with that scheme, while the forward-backward version gives a slight deviation in the computation of u^1.

We can redo the derivation of the initial conditions when $u'(0) = V$:

$$v(0) \approx \frac{1}{2}\left(v^{-\frac{1}{2}} + v^{\frac{1}{2}}\right) = V, \quad \Rightarrow \quad v^{-\frac{1}{2}} = 2V - v^{\frac{1}{2}}\,.$$

Using this $v^{-\frac{1}{2}}$ in

$$v^{\frac{1}{2}} = v^{-\frac{1}{2}} - \Delta t\omega^2 I,$$

then gives $v^{\frac{1}{2}} = V - \frac{1}{2}\Delta t\omega^2 I$. The general initial conditions are therefore

$$u^0 = I, \tag{1.69}$$

$$v^{\frac{1}{2}} = V - \frac{1}{2}\Delta t\omega^2 I\,. \tag{1.70}$$

1.8.2 Implementation of the Scheme on a Staggered Mesh

The algorithm goes like this:

1. Set the initial values (1.69) and (1.70).
2. For $n = 1, 2, \ldots$:
   (a) Compute u^n from (1.68).
   (b) Compute $v^{n+\frac{1}{2}}$ from (1.67).

Implementation with integer indices Translating the schemes (1.68) and (1.67) to computer code faces the problem of how to store and access $v^{n+\frac{1}{2}}$, since arrays only allow integer indices with base 0. We must then introduce a convention: $v^{1+\frac{1}{2}}$ is stored in `v[n]` while $v^{1-\frac{1}{2}}$ is stored in `v[n-1]`. We can then write the algorithm in Python as

```
def solver(I, w, dt, T):
    dt = float(dt)
    Nt = int(round(T/dt))
    u = zeros(Nt+1)
    v = zeros(Nt+1)
    t = linspace(0, Nt*dt, Nt+1)  # mesh for u
    t_v = t + dt/2                # mesh for v

    u[0] = I
    v[0] = 0 - 0.5*dt*w**2*u[0]
    for n in range(1, Nt+1):
        u[n] = u[n-1] + dt*v[n-1]
        v[n] = v[n-1] - dt*w**2*u[n]
    return u, t, v, t_v
```

Note that u and v are returned together with the mesh points such that the complete mesh function for u is described by `u` and `t`, while `v` and `t_v` represent the mesh function for v.

Implementation with half-integer indices Some prefer to see a closer relationship between the code and the mathematics for the quantities with half-integer indices. For example, we would like to replace the updating equation for `v[n]` by

```
v[n+half] = v[n-half] - dt*w**2*u[n]
```

This is easy to do if we could be sure that `n+half` means `n` and `n-half` means `n-1`. A possible solution is to define `half` as a special object such that an integer plus `half` results in the integer, while an integer minus `half` equals the integer minus 1. A simple Python class may realize the `half` object:

```
class HalfInt:
    def __radd__(self, other):
        return other

    def __rsub__(self, other):
        return other - 1

half = HalfInt()
```

The `__radd__` function is invoked for all expressions `n+half` ("right add" with `self` as `half` and `other` as `n`). Similarly, the `__rsub__` function is invoked for `n-half` and results in `n-1`.

Using the `half` object, we can implement the algorithms in an even more readable way:

```
def solver(I, w, dt, T):
    """
    Solve u'=v, v' = - w**2*u for t in (0,T], u(0)=I and v(0)=0,
    by a central finite difference method with time step dt on
    a staggered mesh with v as unknown at (i+1/2)*dt time points.
    """
    dt = float(dt)
    Nt = int(round(T/dt))
    u = zeros(Nt+1)
    v = zeros(Nt+1)
    t = linspace(0, Nt*dt, Nt+1)  # mesh for u
    t_v = t + dt/2                # mesh for v

    u[0] = I
    v[0+half] = 0 - 0.5*dt*w**2*u[0]
    for n in range(1, Nt+1):
        u[n] = u[n-1] + dt*v[n-half]
        v[n+half] = v[n-half] - dt*w**2*u[n]
    return u, t, v[:-1], t_v[:-1]
```

Verification of this code is easy as we can just compare the computed `u` with the `u` produced by the `solver` function in `vib_undamped.py` (which solves $u'' + \omega^2 u = 0$ directly). The values should coincide to machine precision since the two numerical methods are mathematically equivalent. We refer to the file `vib_undamped_staggered.py` for the details of a unit test (`test_staggered`) that checks this property.

1.9 Exercises and Problems

Problem 1.1: Use linear/quadratic functions for verification
Consider the ODE problem

$$u'' + \omega^2 u = f(t), \quad u(0) = I,\ u'(0) = V,\ t \in (0, T]\,.$$

a) Discretize this equation according to $[D_t D_t u + \omega^2 u = f]^n$ and derive the equation for the first time step (u^1).
b) For verification purposes, we use the method of manufactured solutions (MMS) with the choice of $u_e(t) = ct + d$. Find restrictions on c and d from the initial conditions. Compute the corresponding source term f. Show that $[D_t D_t t]^n = 0$ and use the fact that the $D_t D_t$ operator is linear, $[D_t D_t (ct + d)]^n = c[D_t D_t t]^n + [D_t D_t d]^n = 0$, to show that u_e is also a perfect solution of the discrete equations.
c) Use `sympy` to do the symbolic calculations above. Here is a sketch of the program `vib_undamped_verify_mms.py`:

```
import sympy as sym
V, t, I, w, dt = sym.symbols('V t I w dt')  # global symbols
f = None  # global variable for the source term in the ODE

def ode_source_term(u):
    """Return the terms in the ODE that the source term
    must balance, here u'' + w**2*u.
    u is symbolic Python function of t."""
    return sym.diff(u(t), t, t) + w**2*u(t)

def residual_discrete_eq(u):
    """Return the residual of the discrete eq. with u inserted."""
    R = ...
    return sym.simplify(R)

def residual_discrete_eq_step1(u):
    """Return the residual of the discrete eq. at the first
    step with u inserted."""
    R = ...
    return sym.simplify(R)

def DtDt(u, dt):
    """Return 2nd-order finite difference for u_tt.
    u is a symbolic Python function of t.
    """
    return ...
```

```
def main(u):
    """
    Given some chosen solution u (as a function of t, implemented
    as a Python function), use the method of manufactured solutions
    to compute the source term f, and check if u also solves
    the discrete equations.
    """
    print '=== Testing exact solution: %s ===' % u
    print "Initial conditions u(0)=%s, u'(0)=%s:" % \
          (u(t).subs(t, 0), sym.diff(u(t), t).subs(t, 0))

    # Method of manufactured solution requires fitting f
    global f  # source term in the ODE
    f = sym.simplify(ode_lhs(u))

    # Residual in discrete equations (should be 0)
    print 'residual step1:', residual_discrete_eq_step1(u)
    print 'residual:', residual_discrete_eq(u)

def linear():
    main(lambda t: V*t + I)

if __name__ == '__main__':
    linear()
```

Fill in the various functions such that the calls in the `main` function works.

d) The purpose now is to choose a quadratic function $u_e = bt^2 + ct + d$ as exact solution. Extend the `sympy` code above with a function `quadratic` for fitting `f` and checking if the discrete equations are fulfilled. (The function is very similar to `linear`.)
e) Will a polynomial of degree three fulfill the discrete equations?
f) Implement a `solver` function for computing the numerical solution of this problem.
g) Write a test function for checking that the quadratic solution is computed correctly (to machine precision, but the round-off errors accumulate and increase with T) by the `solver` function.

Filename: `vib_undamped_verify_mms`.

Exercise 1.2: Show linear growth of the phase with time
Consider an exact solution $I\cos(\omega t)$ and an approximation $I\cos(\tilde{\omega}t)$. Define the phase error as the time lag between the peak I in the exact solution and the corresponding peak in the approximation after m periods of oscillations. Show that this phase error is linear in m.
Filename: `vib_phase_error_growth`.

Exercise 1.3: Improve the accuracy by adjusting the frequency
According to (1.19), the numerical frequency deviates from the exact frequency by a (dominating) amount $\omega^3\Delta t^2/24 > 0$. Replace the `w` parameter in the algorithm in the `solver` function in `vib_undamped.py` by `w*(1 - (1./24)*w**2*dt**2`

and test how this adjustment in the numerical algorithm improves the accuracy (use $\Delta t = 0.1$ and simulate for 80 periods, with and without adjustment of ω).
Filename: `vib_adjust_w`.

Exercise 1.4: See if adaptive methods improve the phase error
Adaptive methods for solving ODEs aim at adjusting Δt such that the error is within a user-prescribed tolerance. Implement the equation $u'' + u = 0$ in the Odespy[15] software. Use the example from Section 3.2.11 in [9]. Run the scheme with a very low tolerance (say 10^{-14}) and for a long time, check the number of time points in the solver's mesh (`len(solver.t_all)`), and compare the phase error with that produced by the simple finite difference method from Sect. 1.1.2 with the same number of (equally spaced) mesh points. The question is whether it pays off to use an adaptive solver or if equally many points with a simple method gives about the same accuracy.
Filename: `vib_undamped_adaptive`.

Exercise 1.5: Use a Taylor polynomial to compute u^1
As an alternative to computing u^1 by (1.8), one can use a Taylor polynomial with three terms:

$$u(t_1) \approx u(0) + u'(0)\Delta t + \frac{1}{2}u''(0)\Delta t^2 \,.$$

With $u'' = -\omega^2 u$ and $u'(0) = 0$, show that this method also leads to (1.8). Generalize the condition on $u'(0)$ to be $u'(0) = V$ and compute u^1 in this case with both methods.
Filename: `vib_first_step`.

Problem 1.6: Derive and investigate the velocity Verlet method
The velocity Verlet method for $u'' + \omega^2 u = 0$ is based on the following ideas:

1. step u forward from t_n to t_{n+1} using a three-term Taylor series,
2. replace u'' by $-\omega^2 u$
3. discretize $v' = -\omega^2 u$ by a Crank-Nicolson method.

Derive the scheme, implement it, and determine empirically the convergence rate.

Problem 1.7: Find the minimal resolution of an oscillatory function
Sketch the function on a given mesh which has the highest possible frequency. That is, this oscillatory "cos-like" function has its maxima and minima at every two grid points. Find an expression for the frequency of this function, and use the result to find the largest relevant value of $\omega\Delta t$ when ω is the frequency of an oscillating function and Δt is the mesh spacing.
Filename: `vib_largest_wdt`.

[15] https://github.com/hplgit/odespy

Exercise 1.8: Visualize the accuracy of finite differences for a cosine function
We introduce the error fraction

$$E = \frac{[D_t D_t u]^n}{u''(t_n)}$$

to measure the error in the finite difference approximation $D_t D_t u$ to u''. Compute E for the specific choice of a cosine/sine function of the form $u = \exp(i\omega t)$ and show that

$$E = \left(\frac{2}{\omega\Delta t}\right)^2 \sin^2\left(\frac{\omega\Delta t}{2}\right).$$

Plot E as a function of $p = \omega\Delta t$. The relevant values of p are $[0, \pi]$ (see Exercise 1.7 for why $p > \pi$ does not make sense). The deviation of the curve from unity visualizes the error in the approximation. Also expand E as a Taylor polynomial in p up to fourth degree (use, e.g., `sympy`).
Filename: `vib_plot_fd_exp_error`.

Exercise 1.9: Verify convergence rates of the error in energy
We consider the ODE problem $u'' + \omega^2 u = 0$, $u(0) = I$, $u'(0) = V$, for $t \in (0, T]$. The total energy of the solution $E(t) = \frac{1}{2}(u')^2 + \frac{1}{2}\omega^2 u^2$ should stay constant. The error in energy can be computed as explained in Sect. 1.6.

Make a test function in a separate file, where code from `vib_undamped.py` is imported, but the `convergence_rates` and `test_convergence_rates` functions are copied and modified to also incorporate computations of the error in energy and the convergence rate of this error. The expected rate is 2, just as for the solution itself.
Filename: `test_error_conv`.

Exercise 1.10: Use linear/quadratic functions for verification
This exercise is a generalization of Problem 1.1 to the extended model problem (1.71) where the damping term is either linear or quadratic. Solve the various subproblems and see how the results and problem settings change with the generalized ODE in case of linear or quadratic damping. By modifying the code from Problem 1.1, `sympy` will do most of the work required to analyze the generalized problem.
Filename: `vib_verify_mms`.

Exercise 1.11: Use an exact discrete solution for verification
Write a test function in a separate file that employs the exact discrete solution (1.20) to verify the implementation of the `solver` function in the file `vib_undamped.py`.
Filename: `test_vib_undamped_exact_discrete_sol`.

Exercise 1.12: Use analytical solution for convergence rate tests
The purpose of this exercise is to perform convergence tests of the problem (1.71) when $s(u) = cu$, $F(t) = A\sin\phi t$ and there is no damping. Find the complete analytical solution to the problem in this case (most textbooks on mechanics or ordinary differential equations list the various elements you need to write down the exact solution, or you can use symbolic tools like `sympy` or `wolframalpha.com`).

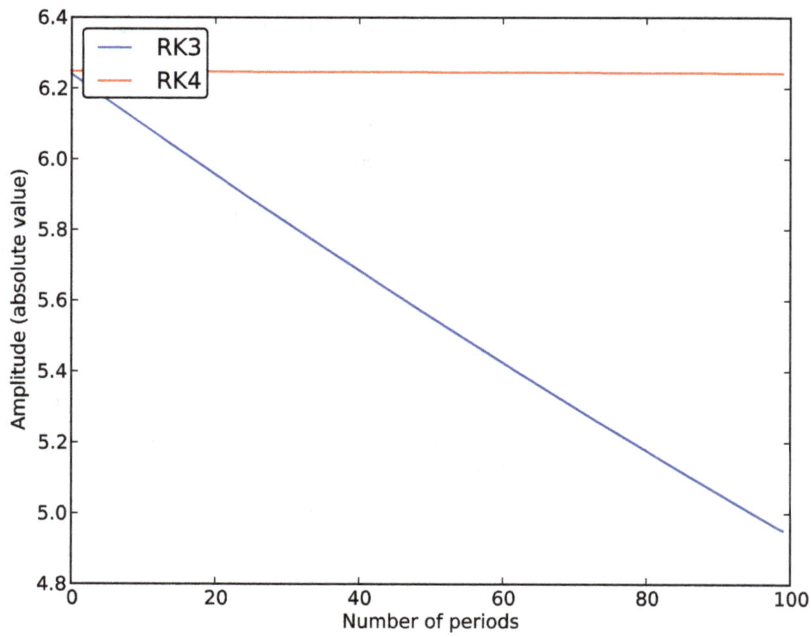

Fig. 1.13 The amplitude as it changes over 100 periods for RK3 and RK4

Modify the `convergence_rate` function from the `vib_undamped.py` program to perform experiments with the extended model. Verify that the error is of order Δt^2.
Filename: `vib_conv_rate`.

Exercise 1.13: Investigate the amplitude errors of many solvers
Use the program `vib_undamped_odespy.py` from Sect. 1.5.4 (utilize the function `amplitudes`) to investigate how well famous methods for 1st-order ODEs can preserve the amplitude of u in undamped oscillations. Test, for example, the 3rd- and 4th-order Runge-Kutta methods (`RK3`, `RK4`), the Crank-Nicolson method (`CrankNicolson`), the 2nd- and 3rd-order Adams-Bashforth methods (`AdamsBashforth2`, `AdamsBashforth3`), and a 2nd-order Backwards scheme (`Backward2Step`). The relevant governing equations are listed in the beginning of Sect. 1.5.

Running the code, we get the plots seen in Fig. 1.13, 1.14, and 1.15. They show that `RK4` is superior to the others, but that also `CrankNicolson` performs well. In fact, with `RK4` the amplitude changes by less than 0.1 per cent over the interval.
Filename: `vib_amplitude_errors`.

Problem 1.14: Minimize memory usage of a simple vibration solver
We consider the model problem $u'' + \omega^2 u = 0$, $u(0) = I$, $u'(0) = V$, solved by a second-order finite difference scheme. A standard implementation typically employs an array `u` for storing all the u^n values. However, at some time level `n+1` where we want to compute `u[n+1]`, all we need of previous `u` values are from level `n` and `n-1`. We can therefore avoid storing the entire array `u`, and instead work with `u[n+1]`, `u[n]`, and `u[n-1]`, named as `u`, `u_n`, `u_nmp1`, for instance. Another possible naming convention is `u`, `u_n[0]`, `u_n[-1]`. Store the solution in a file

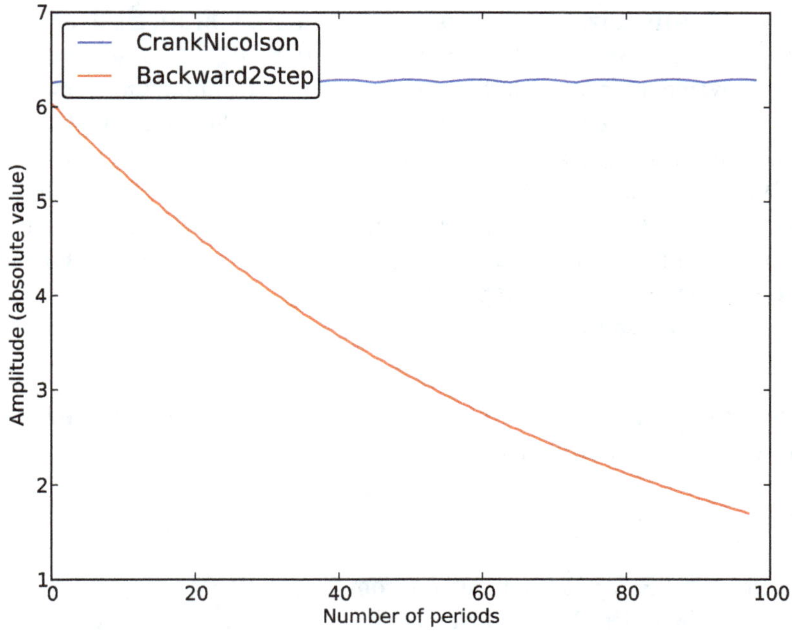

Fig. 1.14 The amplitude as it changes over 100 periods for Crank-Nicolson and Backward 2 step

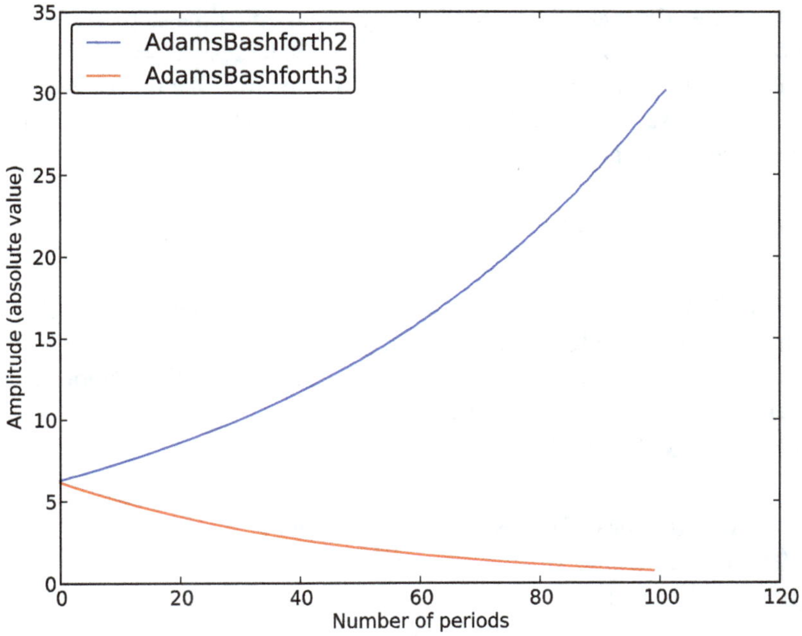

Fig. 1.15 The amplitude as it changes over 100 periods for Adams-Bashforth 2 and 3

for later visualization. Make a test function that verifies the implementation by comparing with the another code for the same problem.
Filename: `vib_memsave0`.

Problem 1.15: Minimize memory usage of a general vibration solver
The program `vib.py` stores the complete solution $u^0, u^1, \ldots, u^{N_t}$ in memory, which is convenient for later plotting. Make a memory minimizing version of this program where only the last three u^{n+1}, u^n, and u^{n-1} values are stored in memory under the names `u`, `u_n`, and `u_nm1` (this is the naming convention used in this book). Write each computed (t_{n+1}, u^{n+1}) pair to file. Visualize the data in the file (a cool solution is to read one line at a time and plot the u value using the line-by-line plotter in the `visualize_front_ascii` function - this technique makes it trivial to visualize very long time simulations).
Filename: `vib_memsave`.

Exercise 1.16: Implement the Euler-Cromer scheme for the generalized model
We consider the generalized model problem

$$mu'' + f(u') + s(u) = F(t), \quad u(0) = I,\ u'(0) = V\,.$$

a) Implement the Euler-Cromer method from Sect. 1.10.8.
b) We expect the Euler-Cromer method to have first-order convergence rate. Make a unit test based on this expectation.
c) Consider a system with $m = 4$, $f(v) = b|v|v$, $b = 0.2$, $s = 2u$, $F = 0$. Compute the solution using the centered difference scheme from Sect. 1.10.1 and the Euler-Cromer scheme for the longest possible time step Δt. We can use the result from the case without damping, i.e., the largest $\Delta t = 2/\omega$, $\omega \approx \sqrt{0.5}$ in this case, but since b will modify the frequency, we take the longest possible time step as a safety factor 0.9 times $2/\omega$. Refine Δt three times by a factor of two and compare the two curves.

Filename: `vib_EulerCromer`.

Problem 1.17: Interpret $[D_tD_tu]^n$ as a forward-backward difference
Show that the difference $[D_tD_tu]^n$ is equal to $[D_t^+D_t^-u]^n$ and $D_t^-D_t^+u]^n$. That is, instead of applying a centered difference twice one can alternatively apply a mixture of forward and backward differences.
Filename: `vib_DtDt_fw_bw`.

Exercise 1.18: Analysis of the Euler-Cromer scheme
The Euler-Cromer scheme for the model problem $u'' + \omega^2 u = 0$, $u(0) = I$, $u'(0) = 0$, is given in (1.55)–(1.54). Find the exact discrete solutions of this scheme and show that the solution for u^n coincides with that found in Sect. 1.4.

Hint Use an "ansatz" $u^n = I \exp(i\tilde{\omega}\Delta t\, n)$ and $v^n = qu^n$, where $\tilde{\omega}$ and q are unknown parameters. The following formula is handy:

$$e^{i\tilde{\omega}\Delta t} + e^{i\tilde{\omega}(-\Delta t)} - 2 = 2\left(\cosh(i\tilde{\omega}\Delta t) - 1\right) = -4\sin^2\left(\frac{\tilde{\omega}\Delta t}{2}\right).$$

1.10 Generalization: Damping, Nonlinearities, and Excitation

We shall now generalize the simple model problem from Sect. 1.1 to include a possibly nonlinear damping term $f(u')$, a possibly nonlinear spring (or restoring) force $s(u)$, and some external excitation $F(t)$:

$$mu'' + f(u') + s(u) = F(t), \quad u(0) = I,\ u'(0) = V,\ t \in (0, T]\,. \qquad (1.71)$$

We have also included a possibly nonzero initial value for $u'(0)$. The parameters m, $f(u')$, $s(u)$, $F(t)$, I, V, and T are input data.

There are two main types of damping (friction) forces: linear $f(u') = bu$, or quadratic $f(u') = bu'|u'|$. Spring systems often feature linear damping, while air resistance usually gives rise to quadratic damping. Spring forces are often linear: $s(u) = cu$, but nonlinear versions are also common, the most famous is the gravity force on a pendulum that acts as a spring with $s(u) \sim \sin(u)$.

1.10.1 A Centered Scheme for Linear Damping

Sampling (1.71) at a mesh point t_n, replacing $u''(t_n)$ by $[D_tD_tu]^n$, and $u'(t_n)$ by $[D_{2t}u]^n$ results in the discretization

$$[mD_tD_tu + f(D_{2t}u) + s(u) = F]^n, \qquad (1.72)$$

which written out means

$$m\frac{u^{n+1} - 2u^n + u^{n-1}}{\Delta t^2} + f\left(\frac{u^{n+1} - u^{n-1}}{2\Delta t}\right) + s(u^n) = F^n, \qquad (1.73)$$

where F^n as usual means $F(t)$ evaluated at $t = t_n$. Solving (1.73) with respect to the unknown u^{n+1} gives a problem: the u^{n+1} inside the f function makes the equation *nonlinear* unless $f(u')$ is a linear function, $f(u') = bu'$. For now we shall assume that f is linear in u'. Then

$$m\frac{u^{n+1} - 2u^n + u^{n-1}}{\Delta t^2} + b\frac{u^{n+1} - u^{n-1}}{2\Delta t} + s(u^n) = F^n, \qquad (1.74)$$

which gives an explicit formula for u at each new time level:

$$u^{n+1} = \left(2mu^n + \left(\frac{b}{2}\Delta t - m\right)u^{n-1} + \Delta t^2(F^n - s(u^n))\right)\left(m + \frac{b}{2}\Delta t\right)^{-1}. \qquad (1.75)$$

For the first time step we need to discretize $u'(0) = V$ as $[D_{2t}u = V]^0$ and combine with (1.75) for $n = 0$. The discretized initial condition leads to

$$u^{-1} = u^1 - 2\Delta t V, \qquad (1.76)$$

which inserted in (1.75) for $n = 0$ gives an equation that can be solved for u^1:

$$u^1 = u^0 + \Delta t\, V + \frac{\Delta t^2}{2m}(-bV - s(u^0) + F^0)\,. \qquad (1.77)$$

1.10.2 A Centered Scheme for Quadratic Damping

When $f(u') = bu'|u'|$, we get a quadratic equation for u^{n+1} in (1.73). This equation can be straightforwardly solved by the well-known formula for the roots of a quadratic equation. However, we can also avoid the nonlinearity by introducing an approximation with an error of order no higher than what we already have from replacing derivatives with finite differences.

We start with (1.71) and only replace u'' by D_tD_tu, resulting in

$$[mD_tD_tu + bu'|u'| + s(u) = F]^n\,. \tag{1.78}$$

Here, $u'|u'|$ is to be computed at time t_n. The idea is now to introduce a *geometric mean*, defined by

$$(w^2)^n \approx w^{n-\frac{1}{2}}w^{n+\frac{1}{2}},$$

for some quantity w depending on time. The error in the geometric mean approximation is $\mathcal{O}(\Delta t^2)$, the same as in the approximation $u'' \approx D_tD_tu$. With $w = u'$ it follows that

$$[u'|u'|]^n \approx u'(t_{n+\frac{1}{2}})|u'(t_{n-\frac{1}{2}})|\,.$$

The next step is to approximate u' at $t_{n\pm 1/2}$, and fortunately a centered difference fits perfectly into the formulas since it involves u values at the mesh points only. With the approximations

$$u'(t_{n+1/2}) \approx [D_tu]^{n+\frac{1}{2}}, \quad u'(t_{n-1/2}) \approx [D_tu]^{n-\frac{1}{2}}, \tag{1.79}$$

we get

$$[u'|u'|]^n \approx [D_tu]^{n+\frac{1}{2}}|[D_tu]^{n-\frac{1}{2}}| = \frac{u^{n+1}-u^n}{\Delta t}\frac{|u^n-u^{n-1}|}{\Delta t}\,. \tag{1.80}$$

The counterpart to (1.73) is then

$$m\frac{u^{n+1}-2u^n+u^{n-1}}{\Delta t^2} + b\frac{u^{n+1}-u^n}{\Delta t}\frac{|u^n-u^{n-1}|}{\Delta t} + s(u^n) = F^n, \tag{1.81}$$

which is linear in the unknown u^{n+1}. Therefore, we can easily solve (1.81) with respect to u^{n+1} and achieve the explicit updating formula

$$\begin{aligned} u^{n+1} &= \left(m + b|u^n - u^{n-1}|\right)^{-1} \\ &\quad \times \left(2mu^n - mu^{n-1} + bu^n|u^n - u^{n-1}| + \Delta t^2(F^n - s(u^n))\right)\,. \end{aligned} \tag{1.82}$$

In the derivation of a special equation for the first time step we run into some trouble: inserting (1.76) in (1.82) for $n = 0$ results in a complicated nonlinear equation for u^1. By thinking differently about the problem we can easily get away with the nonlinearity again. We have for $n = 0$ that $b[u'|u'|]^0 = bV|V|$. Using this value in (1.78) gives

$$[mD_tD_tu + bV|V| + s(u) = F]^0\,. \tag{1.83}$$

Writing this equation out and using (1.76) results in the special equation for the first time step:

$$u^1 = u^0 + \Delta t V + \frac{\Delta t^2}{2m}\left(-bV|V| - s(u^0) + F^0\right). \tag{1.84}$$

1.10.3 A Forward-Backward Discretization of the Quadratic Damping Term

The previous section first proposed to discretize the quadratic damping term $|u'|u'$ using centered differences: $[|D_{2t}|D_{2t}u]^n$. As this gives rise to a nonlinearity in u^{n+1}, it was instead proposed to use a geometric mean combined with centered differences. But there are other alternatives. To get rid of the nonlinearity in $[|D_{2t}|D_{2t}u]^n$, one can think differently: apply a backward difference to $|u'|$, such that the term involves known values, and apply a forward difference to u' to make the term linear in the unknown u^{n+1}. With mathematics,

$$[\beta|u'|u']^n \approx \beta|[D_t^- u]^n|[D_t^+ u]^n = \beta\left|\frac{u^n - u^{n-1}}{\Delta t}\right|\frac{u^{n+1} - u^n}{\Delta t}. \tag{1.85}$$

The forward and backward differences both have an error proportional to Δt so one may think the discretization above leads to a first-order scheme. However, by looking at the formulas, we realize that the forward-backward differences in (1.85) result in exactly the same scheme as in (1.81) where we used a geometric mean and centered differences and committed errors of size $\mathcal{O}(\Delta t^2)$. Therefore, the forward-backward differences in (1.85) act in a symmetric way and actually produce a second-order accurate discretization of the quadratic damping term.

1.10.4 Implementation

The algorithm arising from the methods in Sect.s 1.10.1 and 1.10.2 is very similar to the undamped case in Sect. 1.1.2. The difference is basically a question of different formulas for u^1 and u^{n+1}. This is actually quite remarkable. The equation (1.71) is normally impossible to solve by pen and paper, but possible for some special choices of F, s, and f. On the contrary, the complexity of the nonlinear generalized model (1.71) versus the simple undamped model is not a big deal when we solve the problem numerically!

The computational algorithm takes the form

1. $u^0 = I$
2. compute u^1 from (1.77) if linear damping or (1.84) if quadratic damping
3. for $n = 1, 2, \ldots, N_t - 1$:
   (a) compute u^{n+1} from (1.75) if linear damping or (1.82) if quadratic damping

Modifying the `solver` function for the undamped case is fairly easy, the big difference being many more terms and if tests on the type of damping:

```
def solver(I, V, m, b, s, F, dt, T, damping='linear'):
    """
    Solve m*u'' + f(u') + s(u) = F(t) for t in (0,T],
    u(0)=I and u'(0)=V,
    by a central finite difference method with time step dt.
    If damping is 'linear', f(u')=b*u, while if damping is
    'quadratic', f(u')=b*u'*abs(u').
    F(t) and s(u) are Python functions.
    """
    dt = float(dt); b = float(b); m = float(m) # avoid integer div.
    Nt = int(round(T/dt))
    u = np.zeros(Nt+1)
    t = np.linspace(0, Nt*dt, Nt+1)

    u[0] = I
    if damping == 'linear':
        u[1] = u[0] + dt*V + dt**2/(2*m)*(-b*V - s(u[0]) + F(t[0]))
    elif damping == 'quadratic':
        u[1] = u[0] + dt*V + \
               dt**2/(2*m)*(-b*V*abs(V) - s(u[0]) + F(t[0]))

    for n in range(1, Nt):
        if damping == 'linear':
            u[n+1] = (2*m*u[n] + (b*dt/2 - m)*u[n-1] +
                      dt**2*(F(t[n]) - s(u[n])))/(m + b*dt/2)
        elif damping == 'quadratic':
            u[n+1] = (2*m*u[n] - m*u[n-1] + b*u[n]*abs(u[n] - u[n-1])
                      + dt**2*(F(t[n]) - s(u[n])))/\
                      (m + b*abs(u[n] - u[n-1]))
    return u, t
```

The complete code resides in the file `vib.py`.

1.10.5 Verification

Constant solution For debugging and initial verification, a constant solution is often very useful. We choose $u_e(t) = I$, which implies $V = 0$. Inserted in the ODE, we get $F(t) = s(I)$ for any choice of f. Since the discrete derivative of a constant vanishes (in particular, $[D_{2t}I]^n = 0$, $[D_tI]^n = 0$, and $[D_tD_tI]^n = 0$), the constant solution also fulfills the discrete equations. The constant should therefore be reproduced to machine precision. The function `test_constant` in `vib.py` implements this test.

Linear solution Now we choose a linear solution: $u_e = ct + d$. The initial condition $u(0) = I$ implies $d = I$, and $u'(0) = V$ forces c to be V. Inserting $u_e = Vt + I$ in the ODE with linear damping results in

$$0 + bV + s(Vt + I) = F(t),$$

while quadratic damping requires the source term

$$0 + b|V|V + s(Vt + I) = F(t).$$

Since the finite difference approximations used to compute u' all are exact for a linear function, it turns out that the linear u_e is also a solution of the discrete equations. Exercise 1.10 asks you to carry out all the details.

Quadratic solution Choosing $u_e = bt^2 + Vt + I$, with b arbitrary, fulfills the initial conditions and fits the ODE if F is adjusted properly. The solution also solves the discrete equations with linear damping. However, this quadratic polynomial in t does not fulfill the discrete equations in case of quadratic damping, because the geometric mean used in the approximation of this term introduces an error. Doing Exercise 1.10 will reveal the details. One can fit F^n in the discrete equations such that the quadratic polynomial is reproduced by the numerical method (to machine precision).

Catching bugs How good are the constant and quadratic solutions at catching bugs in the implementation? Let us check that by introducing some bugs.

- Use `m` instead of `2*m` in the denominator of `u[1]`: code works for constant solution, but fails (as it should) for a quadratic one.
- Use `b*dt` instead of `b*dt/2` in the updating formula for `u[n+1]` in case of linear damping: constant and quadratic both fail.
- Use `F[n+1]` instead of `F[n]` in case of linear or quadratic damping: constant solution works, quadratic fails.

We realize that the constant solution is very useful for catching certain bugs because of its simplicity (easy to predict what the different terms in the formula should evaluate to), while the quadratic solution seems capable of detecting all (?) other kinds of typos in the scheme. These results demonstrate why we focus so much on exact, simple polynomial solutions of the numerical schemes in these writings.

1.10.6 Visualization

The functions for visualizations differ significantly from those in the undamped case in the `vib_undamped.py` program because, in the present general case, we do not have an exact solution to include in the plots. Moreover, we have no good estimate of the periods of the oscillations as there will be one period determined by the system parameters, essentially the approximate frequency $\sqrt{s'(0)/m}$ for linear s and small damping, and one period dictated by $F(t)$ in case the excitation is periodic. This is, however, nothing that the program can depend on or make use of. Therefore, the user has to specify T and the window width to get a plot that moves with the graph and shows the most recent parts of it in long time simulations.

The `vib.py` code contains several functions for analyzing the time series signal and for visualizing the solutions.

1.10.7 User Interface

The `main` function is changed substantially from the `vib_undamped.py` code, since we need to specify the new data c, $s(u)$, and $F(t)$. In addition, we must set T and the plot window width (instead of the number of periods we want to simulate as in `vib_undamped.py`). To figure out whether we can use one plot for the whole time series or if we should follow the most recent part of u, we can use the `plot_empricial_freq_and_amplitude` function's estimate of the number of local maxima. This number is now returned from the function and used in `main` to decide on the visualization technique.

```
def main():
    import argparse
    parser = argparse.ArgumentParser()
    parser.add_argument('--I', type=float, default=1.0)
    parser.add_argument('--V', type=float, default=0.0)
    parser.add_argument('--m', type=float, default=1.0)
    parser.add_argument('--c', type=float, default=0.0)
    parser.add_argument('--s', type=str, default='u')
    parser.add_argument('--F', type=str, default='0')
    parser.add_argument('--dt', type=float, default=0.05)
    parser.add_argument('--T', type=float, default=140)
    parser.add_argument('--damping', type=str, default='linear')
    parser.add_argument('--window_width', type=float, default=30)
    parser.add_argument('--savefig', action='store_true')
    a = parser.parse_args()
    from scitools.std import StringFunction
    s = StringFunction(a.s, independent_variable='u')
    F = StringFunction(a.F, independent_variable='t')
    I, V, m, c, dt, T, window_width, savefig, damping = \

       a.I, a.V, a.m, a.c, a.dt, a.T, a.window_width, a.savefig, \
       a.damping

    u, t = solver(I, V, m, c, s, F, dt, T)
    num_periods = empirical_freq_and_amplitude(u, t)
    if num_periods <= 15:
        figure()
        visualize(u, t)
    else:
        visualize_front(u, t, window_width, savefig)
    show()
```

The program `vib.py` contains the above code snippets and can solve the model problem (1.71). As a demo of `vib.py`, we consider the case $I = 1$, $V = 0$, $m = 1$, $c = 0.03$, $s(u) = \sin(u)$, $F(t) = 3\cos(4t)$, $\Delta t = 0.05$, and $T = 140$. The relevant command to run is

Terminal

```
Terminal> python vib.py --s 'sin(u)' --F '3*cos(4*t)' --c 0.03
```

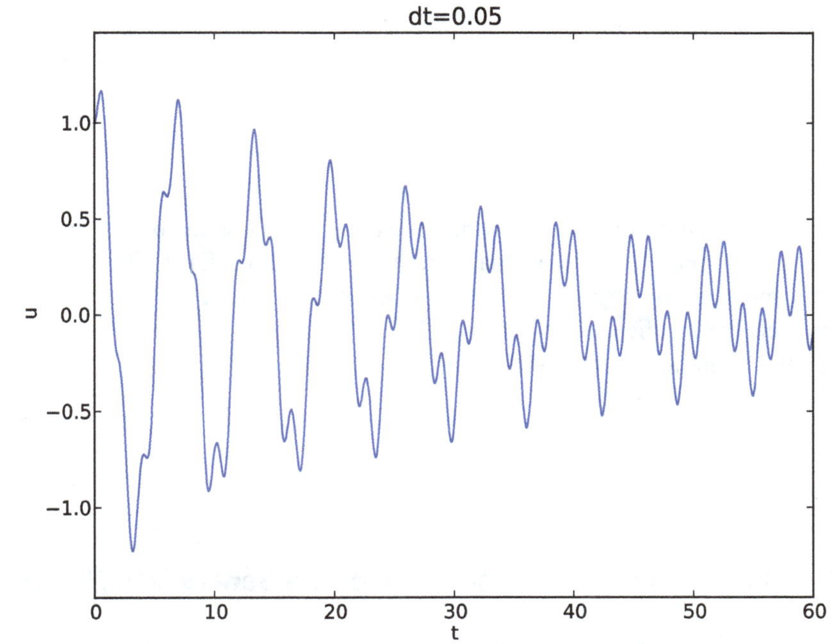

Fig. 1.16 Damped oscillator excited by a sinusoidal function

This results in a moving window following the function[16] on the screen. Figure 1.16 shows a part of the time series.

1.10.8 The Euler-Cromer Scheme for the Generalized Model

The ideas of the Euler-Cromer method from Sect. 1.7 carry over to the generalized model. We write (1.71) as two equations for u and $v = u'$. The first equation is taken as the one with v' on the left-hand side:

$$v' = \frac{1}{m}(F(t) - s(u) - f(v)), \tag{1.86}$$

$$u' = v\,. \tag{1.87}$$

Again, the idea is to step (1.86) forward using a standard Forward Euler method, while we update u from (1.87) with a Backward Euler method, utilizing the recent, computed v^{n+1} value. In detail,

$$\frac{v^{n+1} - v^n}{\Delta t} = \frac{1}{m}(F(t_n) - s(u^n) - f(v^n)), \tag{1.88}$$

$$\frac{u^{n+1} - u^n}{\Delta t} = v^{n+1}, \tag{1.89}$$

[16] http://tinyurl.com/hbcasmj/vib/html//mov-vib/vib_generalized_dt0.05/index.html

resulting in the explicit scheme

$$v^{n+1} = v^n + \Delta t \frac{1}{m}(F(t_n) - s(u^n) - f(v^n)), \tag{1.90}$$

$$u^{n+1} = u^n + \Delta t\, v^{n+1}\,. \tag{1.91}$$

We immediately note one very favorable feature of this scheme: all the nonlinearities in $s(u)$ and $f(v)$ are evaluated at a previous time level. This makes the Euler-Cromer method easier to apply and hence much more convenient than the centered scheme for the second-order ODE (1.71).

The initial conditions are trivially set as

$$v^0 = V, \tag{1.92}$$

$$u^0 = I\,. \tag{1.93}$$

1.10.9 The Störmer-Verlet Algorithm for the Generalized Model

We can easily apply the ideas from Sect. 1.7.4 to extend that method to the generalized model

$$v' = \frac{1}{m}(F(t) - s(u) - f(v)),$$

$$u' = v\,.$$

However, since the scheme is essentially centered differences for the ODE system on a staggered mesh, we do not go into detail here, but refer to Sect. 1.10.10.

1.10.10 A Staggered Euler-Cromer Scheme for a Generalized Model

The more general model for vibration problems,

$$mu'' + f(u') + s(u) = F(t), \quad u(0) = I,\ u'(0) = V,\ t \in (0, T], \tag{1.94}$$

can be rewritten as a first-order ODE system

$$v' = m^{-1}\left(F(t) - f(v) - s(u)\right), \tag{1.95}$$

$$u' = v\,. \tag{1.96}$$

It is natural to introduce a staggered mesh (see Sect. 1.8.1) and seek u at mesh points t_n (the numerical value is denoted by u^n) and v between mesh points at $t_{n+1/2}$ (the numerical value is denoted by $v^{n+\frac{1}{2}}$). A centered difference approximation to (1.96)–(1.95) can then be written in operator notation as

$$[D_t v = m^{-1}\left(F(t) - f(v) - s(u)\right)]^n, \tag{1.97}$$

$$[D_t u = v]^{n+\frac{1}{2}}\,. \tag{1.98}$$

Written out,

$$\frac{v^{n+\frac{1}{2}} - v^{n-\frac{1}{2}}}{\Delta t} = m^{-1}\left(F^n - f(v^n) - s(u^n)\right), \tag{1.99}$$

$$\frac{u^n - u^{n-1}}{\Delta t} = v^{n+\frac{1}{2}}. \tag{1.100}$$

With linear damping, $f(v) = bv$, we can use an arithmetic mean for $f(v^n)$: $f(v^n) \approx= \frac{1}{2}(f(v^{n-\frac{1}{2}}) + f(v^{n+\frac{1}{2}}))$. The system (1.99)–(1.100) can then be solved with respect to the unknowns u^n and $v^{n+\frac{1}{2}}$:

$$v^{n+\frac{1}{2}} = \left(1 + \frac{b}{2m}\Delta t\right)^{-1}\left(v^{n-\frac{1}{2}} + \Delta t m^{-1}\left(F^n - \frac{1}{2}f(v^{n-\frac{1}{2}}) - s(u^n)\right)\right), \tag{1.101}$$

$$u^n = u^{n-1} + \Delta t v^{n-\frac{1}{2}}. \tag{1.102}$$

In case of quadratic damping, $f(v) = b|v|v$, we can use a geometric mean: $f(v^n) \approx b|v^{n-\frac{1}{2}}|v^{n+\frac{1}{2}}$. Inserting this approximation in (1.99)–(1.100) and solving for the unknowns u^n and $v^{n+\frac{1}{2}}$ results in

$$v^{n+\frac{1}{2}} = \left(1 + \frac{b}{m}|v^{n-\frac{1}{2}}|\Delta t\right)^{-1}\left(v^{n-\frac{1}{2}} + \Delta t m^{-1}\left(F^n - s(u^n)\right)\right), \tag{1.103}$$

$$u^n = u^{n-1} + \Delta t v^{n-\frac{1}{2}}. \tag{1.104}$$

The initial conditions are derived at the end of Sect. 1.8.1:

$$u^0 = I, \tag{1.105}$$

$$v^{\frac{1}{2}} = V - \frac{1}{2}\Delta t\omega^2 I. \tag{1.106}$$

1.10.11 The PEFRL 4th-Order Accurate Algorithm

A variant of the Euler-Cromer type of algorithm, which provides an error $\mathcal{O}(\Delta t^4)$ if $f(v) = 0$, is called PEFRL [14]. This algorithm is very well suited for integrating dynamic systems (especially those without damping) over very long time periods. Define

$$g(u, v) = \frac{1}{m}(F(t) - s(u) - f(v)).$$

The algorithm is explicit and features these steps:

$$
\begin{aligned}
u^{n+1,1} &= u^n + \xi \Delta t v^n, && (1.107)\\
v^{n+1,1} &= v^n + \frac{1}{2}(1-2\lambda)\Delta t g(u^{n+1,1}, v^n), && (1.108)\\
u^{n+1,2} &= u^{n+1,1} + \chi \Delta t v^{n+1,1}, && (1.109)\\
v^{n+1,2} &= v^{n+1,1} + \lambda \Delta t g(u^{n+1,2}, v^{n+1,1}), && (1.110)\\
u^{n+1,3} &= u^{n+1,2} + (1-2(\chi+\xi))\Delta t v^{n+1,2}, && (1.111)\\
v^{n+1,3} &= v^{n+1,2} + \lambda \Delta t g(u^{n+1,3}, v^{n+1,2}), && (1.112)\\
u^{n+1,4} &= u^{n+1,3} + \chi \Delta t v^{n+1,3}, && (1.113)\\
v^{n+1} &= v^{n+1,3} + \frac{1}{2}(1-2\lambda)\Delta t g(u^{n+1,4}, v^{n+1,3}), && (1.114)\\
u^{n+1} &= u^{n+1,4} + \xi \Delta t v^{n+1}\,. && (1.115)
\end{aligned}
$$

The parameters ξ, λ, and ξ have the values

$$
\begin{aligned}
\xi &= 0.1786178958448091, && (1.116)\\
\lambda &= -0.2123418310626054, && (1.117)\\
\chi &= -0.06626458266981849\,. && (1.118)
\end{aligned}
$$

1.11 Exercises and Problems

Exercise 1.19: Implement the solver via classes
Reimplement the `vib.py` program using a class `Problem` to hold all the physical parameters of the problem, a class `Solver` to hold the numerical parameters and compute the solution, and a class `Visualizer` to display the solution.

Hint Use the ideas and examples from Sections 5.5.1 and 5.5.2 in [9]. More specifically, make a superclass `Problem` for holding the scalar physical parameters of a problem and let subclasses implement the $s(u)$ and $F(t)$ functions as methods. Try to call up as much existing functionality in `vib.py` as possible.
Filename: `vib_class`.

Problem 1.20: Use a backward difference for the damping term
As an alternative to discretizing the damping terms $\beta u'$ and $\beta |u'|u'$ by centered differences, we may apply backward differences:

$$
\begin{aligned}
[u']^n &\approx [D_t^- u]^n,\\
[|u'|u']^n &\approx [|D_t^- u|D_t^- u]^n\\
&= |[D_t^- u]^n|[D_t^- u]^n\,.
\end{aligned}
$$

The advantage of the backward difference is that the damping term is evaluated using known values u^n and u^{n-1} only. Extend the `vib.py` code with a scheme

based on using backward differences in the damping terms. Add statements to compare the original approach with centered difference and the new idea launched in this exercise. Perform numerical experiments to investigate how much accuracy that is lost by using the backward differences.
Filename: `vib_gen_bwdamping`.

Exercise 1.21: Use the forward-backward scheme with quadratic damping
We consider the generalized model with quadratic damping, expressed as a system of two first-order equations as in Sect. 1.10.10:

$$\begin{aligned} u' &= v, \\ v' &= \frac{1}{m}\left(F(t) - \beta|v|v - s(u)\right) . \end{aligned}$$

However, contrary to what is done in Sect. 1.10.10, we want to apply the idea of a forward-backward discretization: u is marched forward by a one-sided Forward Euler scheme applied to the first equation, and thereafter v can be marched forward by a Backward Euler scheme in the second equation, see in Sect. 1.7. Express the idea in operator notation and write out the scheme. Unfortunately, the backward difference for the v equation creates a nonlinearity $|v^{n+1}|v^{n+1}$. To linearize this nonlinearity, use the known value v^n inside the absolute value factor, i.e., $|v^{n+1}|v^{n+1} \approx |v^n|v^{n+1}$. Show that the resulting scheme is equivalent to the one in Sect. 1.10.10 for some time level $n \geq 1$.

What we learn from this exercise is that the first-order differences and the linearization trick play together in "the right way" such that the scheme is as good as when we (in Sect. 1.10.10) carefully apply centered differences and a geometric mean on a staggered mesh to achieve second-order accuracy. There is a difference in the handling of the initial conditions, though, as explained at the end of Sect. 1.7.
Filename: `vib_gen_bwdamping`.

1.12 Applications of Vibration Models

The following text derives some of the most well-known physical problems that lead to second-order ODE models of the type addressed in this book. We consider a simple spring-mass system; thereafter extended with nonlinear spring, damping, and external excitation; a spring-mass system with sliding friction; a simple and a physical (classical) pendulum; and an elastic pendulum.

1.12.1 Oscillating Mass Attached to a Spring

The most fundamental mechanical vibration system is depicted in Fig. 1.17. A body with mass m is attached to a spring and can move horizontally without friction (in the wheels). The position of the body is given by the vector $\boldsymbol{r}(t) = u(t)\boldsymbol{i}$, where $\boldsymbol{i}$ is a unit vector in x direction. There is only one force acting on the body: a spring force $\boldsymbol{F}_s = -ku\boldsymbol{i}$, where k is a constant. The point $x = 0$, where $u = 0$, must

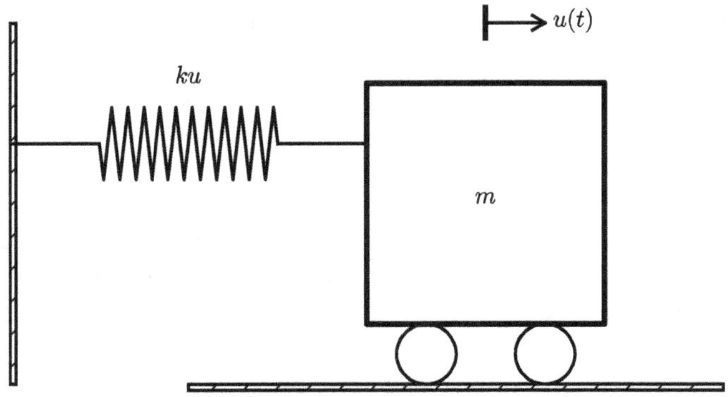

Fig. 1.17 Simple oscillating mass

therefore correspond to the body's position where the spring is neither extended nor compressed, so the force vanishes.

The basic physical principle that governs the motion of the body is Newton's second law of motion: $\boldsymbol{F} = m\boldsymbol{a}$, where $\boldsymbol{F}$ is the sum of forces on the body, m is its mass, and $\boldsymbol{a} = \ddot{\boldsymbol{r}}$ is the acceleration. We use the dot for differentiation with respect to time, which is usual in mechanics. Newton's second law simplifies here to $-\boldsymbol{F}_s = m\ddot{u}\boldsymbol{i}$, which translates to

$$-ku = m\ddot{u}\,.$$

Two initial conditions are needed: $u(0) = I$, $\dot{u}(0) = V$. The ODE problem is normally written as

$$m\ddot{u} + ku = 0, \quad u(0) = I,\ \dot{u}(0) = V\,. \tag{1.119}$$

It is not uncommon to divide by m and introduce the frequency $\omega = \sqrt{k/m}$:

$$\ddot{u} + \omega^2 u = 0, \quad u(0) = I,\ \dot{u}(0) = V\,. \tag{1.120}$$

This is the model problem in the first part of this chapter, with the small difference that we write the time derivative of u with a dot above, while we used u' and u'' in previous parts of the book.

Since only one scalar mathematical quantity, $u(t)$, describes the complete motion, we say that the mechanical system has one degree of freedom (DOF).

Scaling For numerical simulations it is very convenient to scale (1.120) and thereby get rid of the problem of finding relevant values for all the parameters m, k, I, and V. Since the amplitude of the oscillations are dictated by I and V (or more precisely, V/ω), we scale u by I (or V/ω if $I = 0$):

$$\bar{u} = \frac{u}{I}, \quad \bar{t} = \frac{t}{t_c}\,.$$

The time scale t_c is normally chosen as the inverse period $2\pi/\omega$ or angular frequency $1/\omega$, most often as $t_c = 1/\omega$. Inserting the dimensionless quantities $\bar{u}$ and

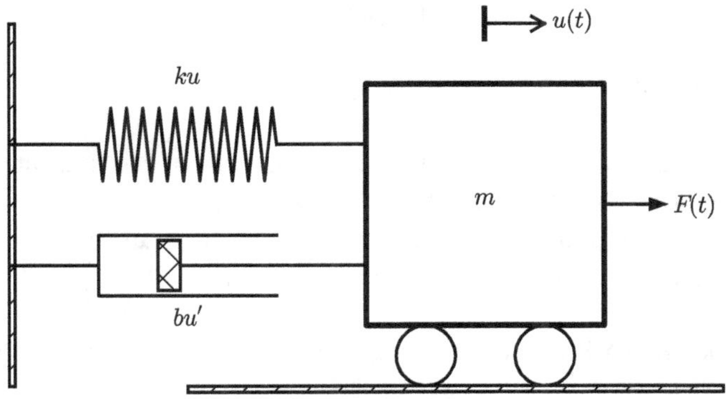

Fig. 1.18 General oscillating system

$\bar{t}$ in (1.120) results in the scaled problem

$$\frac{d^2\bar{u}}{d\bar{t}^2} + \bar{u} = 0, \quad \bar{u}(0) = 1, \ \frac{\bar{u}}{\bar{t}}(0) = \beta = \frac{V}{I\omega},$$

where β is a dimensionless number. Any motion that starts from rest ($V = 0$) is free of parameters in the scaled model!

The physics The typical physics of the system in Fig. 1.17 can be described as follows. Initially, we displace the body to some position I, say at rest ($V = 0$). After releasing the body, the spring, which is extended, will act with a force $-kI\boldsymbol{i}$ and pull the body to the left. This force causes an acceleration and therefore increases velocity. The body passes the point $x = 0$, where $u = 0$, and the spring will then be compressed and act with a force $kx\boldsymbol{i}$ against the motion and cause retardation. At some point, the motion stops and the velocity is zero, before the spring force $kx\boldsymbol{i}$ has worked long enough to push the body in positive direction. The result is that the body accelerates back and forth. As long as there is no friction forces to damp the motion, the oscillations will continue forever.

1.12.2 General Mechanical Vibrating System

The mechanical system in Fig. 1.17 can easily be extended to the more general system in Fig. 1.18, where the body is attached to a spring and a dashpot, and also subject to an environmental force $F(t)\boldsymbol{i}$. The system has still only one degree of freedom since the body can only move back and forth parallel to the x axis. The spring force was linear, $\boldsymbol{F}_s = -ku\boldsymbol{i}$, in Sect. 1.12.1, but in more general cases it can depend nonlinearly on the position. We therefore set $\boldsymbol{F}_s = s(u)\boldsymbol{i}$. The dashpot, which acts as a damper, results in a force $\boldsymbol{F}_d$ that depends on the body's velocity $\dot{u}$ and that always acts against the motion. The mathematical model of the force is written $\boldsymbol{F}_d = f(\dot{u})\boldsymbol{i}$. A positive $\dot{u}$ must result in a force acting in the positive x direction. Finally, we have the external environmental force $\boldsymbol{F}_e = F(t)\boldsymbol{i}$.

Newton's second law of motion now involves three forces:

$$F(t)\boldsymbol{i} - f(\dot{u})\boldsymbol{i} - s(u)\boldsymbol{i} = m\ddot{u}\boldsymbol{i}\,.$$

The common mathematical form of the ODE problem is

$$m\ddot{u} + f(\dot{u}) + s(u) = F(t), \quad u(0) = I, \ \dot{u}(0) = V\,. \tag{1.121}$$

This is the generalized problem treated in the last part of the present chapter, but with prime denoting the derivative instead of the dot.

The most common models for the spring and dashpot are linear: $f(\dot{u}) = b\dot{u}$ with a constant $b \geq 0$, and $s(u) = ku$ for a constant k.

Scaling A specific scaling requires specific choices of f, s, and F. Suppose we have

$$f(\dot{u}) = b|\dot{u}|\dot{u}, \quad s(u) = ku, \quad F(t) = A\sin(\phi t)\,.$$

We introduce dimensionless variables as usual, $\bar{u} = u/u_c$ and $\bar{t} = t/t_c$. The scale u_c depends both on the initial conditions and F, but as time grows, the effect of the initial conditions die out and F will drive the motion. Inserting $\bar{u}$ and $\bar{t}$ in the ODE gives

$$m\frac{u_c}{t_c^2}\frac{d^2\bar{u}}{d\bar{t}^2} + b\frac{u_c^2}{t_c^2}\left|\frac{d\bar{u}}{d\bar{t}}\right|\frac{d\bar{u}}{d\bar{t}} + ku_c\bar{u} = A\sin(\phi t_c\bar{t})\,.$$

We divide by u_c/t_c^2 and demand the coefficients of the $\bar{u}$ and the forcing term from $F(t)$ to have unit coefficients. This leads to the scales

$$t_c = \sqrt{\frac{m}{k}}, \quad u_c = \frac{A}{k}\,.$$

The scaled ODE becomes

$$\frac{d^2\bar{u}}{d\bar{t}^2} + 2\beta\left|\frac{d\bar{u}}{d\bar{t}}\right|\frac{d\bar{u}}{d\bar{t}} + \bar{u} = \sin(\gamma\bar{t}), \tag{1.122}$$

where there are two dimensionless numbers:

$$\beta = \frac{Ab}{2mk}, \quad \gamma = \phi\sqrt{\frac{m}{k}}\,.$$

The β number measures the size of the damping term (relative to unity) and is assumed to be small, basically because b is small. The ϕ number is the ratio of the time scale of free vibrations and the time scale of the forcing. The scaled initial conditions have two other dimensionless numbers as values:

$$\bar{u}(0) = \frac{Ik}{A}, \quad \frac{d\bar{u}}{d\bar{t}} = \frac{t_c}{u_c}V = \frac{V}{A}\sqrt{mk}\,.$$

1.12.3 A Sliding Mass Attached to a Spring

Consider a variant of the oscillating body in Sect. 1.12.1 and Fig. 1.17: the body rests on a flat surface, and there is sliding friction between the body and the surface. Figure 1.19 depicts the problem.

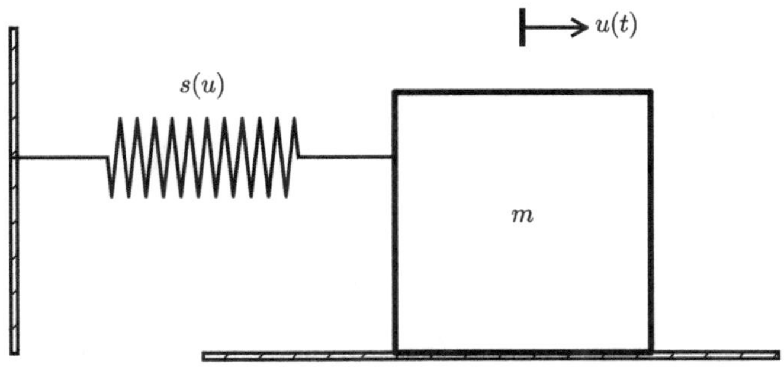

Fig. 1.19 Sketch of a body sliding on a surface

The body is attached to a spring with spring force $-s(u)\boldsymbol{i}$. The friction force is proportional to the normal force on the surface, $-mg\boldsymbol{j}$, and given by $-f(\dot{u})\boldsymbol{i}$, where

$$f(\dot{u}) = \begin{cases} -\mu mg, & \dot{u} < 0, \\ \mu mg, & \dot{u} > 0, \\ 0, & \dot{u} = 0 \end{cases} .$$

Here, μ is a friction coefficient. With the signum function

$$\text{sign(x)} = \begin{cases} -1, & x < 0, \\ 1, & x > 0, \\ 0, & x = 0 \end{cases}$$

we can simply write $f(\dot{u}) = \mu mg\,\text{sign}(\dot{u})$ (the sign function is implemented by `numpy.sign`).

The equation of motion becomes

$$m\ddot{u} + \mu mg\text{sign}(\dot{u}) + s(u) = 0, \quad u(0) = I,\ \dot{u}(0) = V\,. \tag{1.123}$$

1.12.4 A Jumping Washing Machine

A washing machine is placed on four springs with efficient dampers. If the machine contains just a few clothes, the circular motion of the machine induces a sinusoidal external force from the floor and the machine will jump up and down if the frequency of the external force is close to the natural frequency of the machine and its spring-damper system.

1.12.5 Motion of a Pendulum

Simple pendulum A classical problem in mechanics is the motion of a pendulum. We first consider a simplified pendulum[17] (sometimes also called a mathematical

[17] https://en.wikipedia.org/wiki/Pendulum

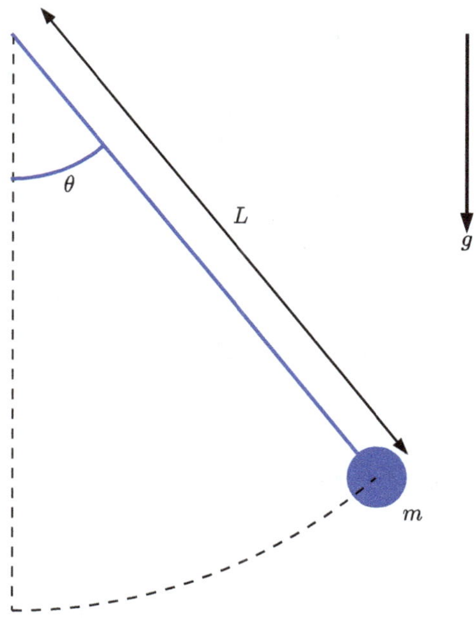

Fig. 1.20 Sketch of a simple pendulum

pendulum): a small body of mass m is attached to a massless wire and can oscillate back and forth in the gravity field. Figure 1.20 shows a sketch of the problem.

The motion is governed by Newton's 2nd law, so we need to find expressions for the forces and the acceleration. Three forces on the body are considered: an unknown force S from the wire, the gravity force mg, and an air resistance force, $\frac{1}{2}C_D\varrho A|v|v$, hereafter called the drag force, directed against the velocity of the body. Here, C_D is a drag coefficient, ϱ is the density of air, A is the cross section area of the body, and v is the magnitude of the velocity.

We introduce a coordinate system with polar coordinates and unit vectors $\boldsymbol{i}_r$ and $\boldsymbol{i}_\theta$ as shown in Fig. 1.21. The position of the center of mass of the body is

$$\boldsymbol{r}(t) = x_0\boldsymbol{i} + y_0\boldsymbol{j} + L\boldsymbol{i}_r,$$

where $\boldsymbol{i}$ and $\boldsymbol{j}$ are unit vectors in the corresponding Cartesian coordinate system in the x and y directions, respectively. We have that $\boldsymbol{i}_r = \cos\theta\boldsymbol{i} + \sin\theta\boldsymbol{j}$.

The forces are now expressed as follows.

- Wire force: $-S\boldsymbol{i}_r$
- Gravity force: $-mg\boldsymbol{j} = mg(-\sin\theta\,\boldsymbol{i}_\theta + \cos\theta\,\boldsymbol{i}_r)$
- Drag force: $-\frac{1}{2}C_D\varrho A|v|v\,\boldsymbol{i}_\theta$

Since a positive velocity means movement in the direction of $\boldsymbol{i}_\theta$, the drag force must be directed along $-\boldsymbol{i}_\theta$ so it works against the motion. We assume motion in air so that the added mass effect can be neglected (for a spherical body, the added mass is $\frac{1}{2}\varrho V$, where V is the volume of the body). Also the buoyancy effect can be neglected for motion in the air when the density difference between the fluid and the body is so significant.

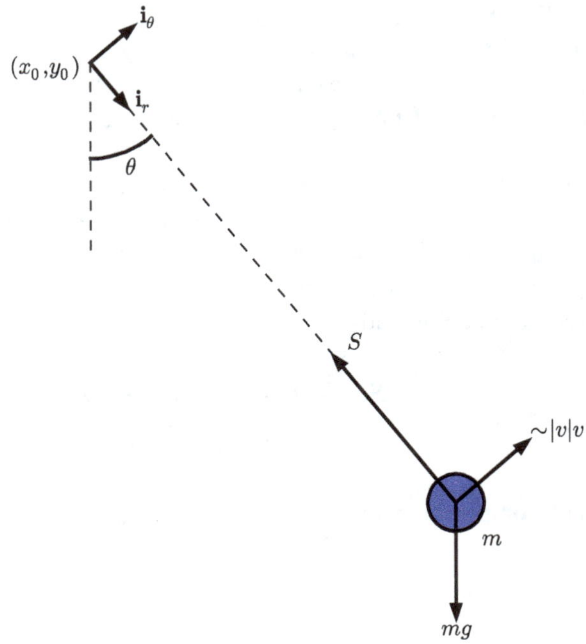

Fig. 1.21 Forces acting on a simple pendulum

The velocity of the body is found from $\boldsymbol{r}$:

$$\boldsymbol{v}(t) = \dot{\boldsymbol{r}}(t) = \frac{d}{d\theta}(x_0\boldsymbol{i} + y_0\boldsymbol{j} + L\boldsymbol{i}_r)\frac{d\theta}{dt} = L\dot{\theta}\boldsymbol{i}_\theta,$$

since $\frac{d}{d\theta}\boldsymbol{i}_r = \boldsymbol{i}_\theta$. It follows that $v = |\boldsymbol{v}| = L\dot{\theta}$. The acceleration is

$$\boldsymbol{a}(t) = \dot{\boldsymbol{v}}(r) = \frac{d}{dt}(L\dot{\theta}\boldsymbol{i}_\theta) = L\ddot{\theta}\boldsymbol{i}_\theta + L\dot{\theta}\frac{d\boldsymbol{i}_\theta}{d\theta}\dot{\theta} = L\ddot{\theta}\boldsymbol{i}_\theta - L\dot{\theta}^2\boldsymbol{i}_r,$$

since $\frac{d}{d\theta}\boldsymbol{i}_\theta = -\boldsymbol{i}_r$.

Newton's 2nd law of motion becomes

$$-S\boldsymbol{i}_r + mg(-\sin\theta\,\boldsymbol{i}_\theta + \cos\theta\,\boldsymbol{i}_r) - \frac{1}{2}C_D\varrho AL^2|\dot{\theta}|\dot{\theta}\,\boldsymbol{i}_\theta = mL\ddot{\theta}\dot{\theta}\,\boldsymbol{i}_\theta - L\dot{\theta}^2\boldsymbol{i}_r,$$

leading to two component equations

$$-S + mg\cos\theta = -L\dot{\theta}^2, \quad (1.124)$$

$$-mg\sin\theta - \frac{1}{2}C_D\varrho AL^2|\dot{\theta}|\dot{\theta} = mL\ddot{\theta}\,. \quad (1.125)$$

From (1.124) we get an expression for $S = mg\cos\theta + L\dot{\theta}^2$, and from (1.125) we get a differential equation for the angle $\theta(t)$. This latter equation is ordered as

$$m\ddot{\theta} + \frac{1}{2}C_D\varrho AL|\dot{\theta}|\dot{\theta} + \frac{mg}{L}\sin\theta = 0\,. \quad (1.126)$$

Two initial conditions are needed: $\theta = \Theta$ and $\dot{\theta} = \Omega$. Normally, the pendulum motion is started from rest, which means $\Omega = 0$.

Equation (1.126) fits the general model used in (1.71) in Sect. 1.10 if we define $u = \theta$, $f(u') = \frac{1}{2}C_D\varrho AL|\dot{u}|\dot{u}$, $s(u) = L^{-1}mg\sin u$, and $F = 0$. If the body is a sphere with radius R, we can take $C_D = 0.4$ and $A = \pi R^2$. Exercise 1.25 asks you to scale the equations and carry out specific simulations with this model.

Physical pendulum The motion of a compound or physical pendulum where the wire is a rod with mass, can be modeled very similarly. The governing equation is $I\boldsymbol{a} = \boldsymbol{T}$ where I is the moment of inertia of the entire body about the point (x_0, y_0), and $\boldsymbol{T}$ is the sum of moments of the forces with respect to (x_0, y_0). The vector equation reads

$$\begin{aligned}&\boldsymbol{r}\times\left(-S\boldsymbol{i}_r + mg(-\sin\theta\boldsymbol{i}_\theta + \cos\theta\boldsymbol{i}_r) - \frac{1}{2}C_D\varrho AL^2|\dot{\theta}|\dot{\theta}\boldsymbol{i}_\theta\right)\\ &= I(L\ddot{\theta}\dot{\theta}\boldsymbol{i}_\theta - L\dot{\theta}^2\boldsymbol{i}_r)\,.\end{aligned}$$

The component equation in $\boldsymbol{i}_\theta$ direction gives the equation of motion for $\theta(t)$:

$$I\ddot{\theta} + \frac{1}{2}C_D\varrho AL^3|\dot{\theta}|\dot{\theta} + mgL\sin\theta = 0\,. \tag{1.127}$$

1.12.6 Dynamic Free Body Diagram During Pendulum Motion

Usually one plots the mathematical quantities as functions of time to visualize the solution of ODE models. Exercise 1.25 asks you to do this for the motion of a pendulum in the previous section. However, sometimes it is more instructive to look at other types of visualizations. For example, we have the pendulum and the free body diagram in Fig. 1.20 and 1.21. We may think of these figures as animations in time instead. Especially the free body diagram will show both the motion of the pendulum *and* the size of the forces during the motion. The present section exemplifies how to make such a dynamic body diagram. Two typical snapshots of free body diagrams are displayed below (the drag force is magnified 5 times to become more visual!).

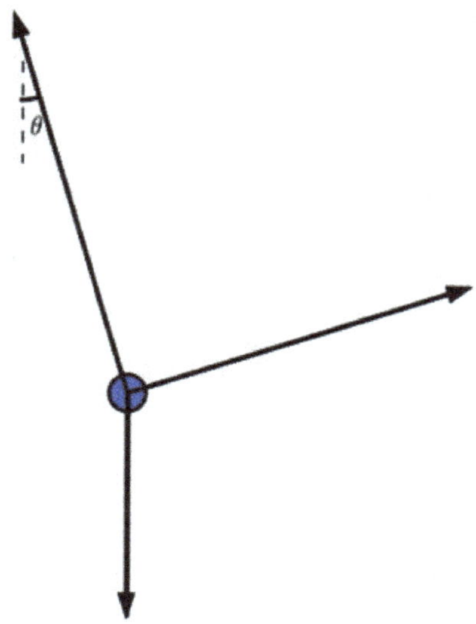

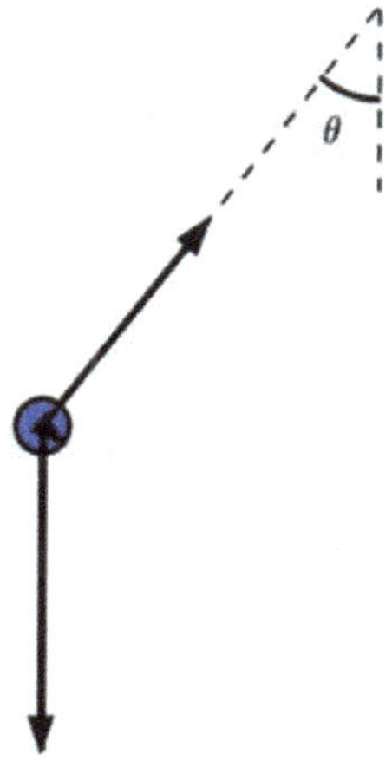

Dynamic physical sketches, coupled to the numerical solution of differential equations, requires a program to produce a sketch for the situation at each time level. Pysketcher[18] is such a tool. In fact (and not surprising!) Fig. 1.20 and 1.21 were drawn using Pysketcher. The details of the drawings are explained in the Pysketcher tutorial[19]. Here, we outline how this type of sketch can be used to create an animated free body diagram during the motion of a pendulum.

Pysketcher is actually a layer of useful abstractions on top of standard plotting packages. This means that we in fact apply Matplotlib to make the animated free body diagram, but instead of dealing with a wealth of detailed Matplotlib commands, we can express the drawing in terms of more high-level objects, e.g., objects for the wire, angle θ, body with mass m, arrows for forces, etc. When the position of these objects are given through variables, we can just couple those variables to the dynamic solution of our ODE and thereby make a unique drawing for each θ value in a simulation.

Writing the solver Let us start with the most familiar part of the current problem: writing the solver function. We use Odespy for this purpose. We also work with dimensionless equations. Since θ can be viewed as dimensionless, we only need to introduce a dimensionless time, here taken as $\bar{t} = t/\sqrt{L/g}$. The resulting dimensionless mathematical model for θ, the dimensionless angular velocity ω, the dimensionless wire force $\bar{S}$, and the dimensionless drag force $\bar{D}$ is then

$$
\begin{aligned}
\frac{d\omega}{d\bar{t}} &= -\alpha|\omega|\omega - \sin\theta, &(1.128)\\
\frac{d\theta}{d\bar{t}} &= \omega, &(1.129)\\
\bar{S} &= \omega^2 + \cos\theta, &(1.130)\\
\bar{D} &= -\alpha|\omega|\omega, &(1.131)
\end{aligned}
$$

with

$$\alpha = \frac{C_D \varrho \pi R^2 L}{2m},$$

as a dimensionless parameter expressing the ratio of the drag force and the gravity force. The dimensionless ω is made non-dimensional by the time, so $\omega\sqrt{L/g}$ is the corresponding angular frequency with dimensions.

A suitable function for computing (1.128)–(1.131) is listed below.

```
def simulate(alpha, Theta, dt, T):
    import odespy

    def f(u, t, alpha):
        omega, theta = u
        return [-alpha*omega*abs(omega) - sin(theta),
                omega]
```

[18] https://github.com/hplgit/pysketcher

[19] http://hplgit.github.io/pysketcher/doc/web/index.html

```
    import numpy as np
    Nt = int(round(T/float(dt)))
    t = np.linspace(0, Nt*dt, Nt+1)
    solver = odespy.RK4(f, f_args=[alpha])
    solver.set_initial_condition([0, Theta])
    u, t = solver.solve(
        t, terminate=lambda u, t, n: abs(u[n,1]) < 1E-3)
    omega = u[:,0]
    theta = u[:,1]
    S = omega**2 + np.cos(theta)
    drag = -alpha*np.abs(omega)*omega
    return t, theta, omega, S, drag
```

Drawing the free body diagram The `sketch` function below applies Pysketcher objects to build a diagram like that in Fig. 1.21, except that we have removed the rotation point (x_0, y_0) and the unit vectors in polar coordinates as these objects are not important for an animated free body diagram.

```
import sys
try:
    from pysketcher import *
except ImportError:
    print 'Pysketcher must be installed from'
    print 'https://github.com/hplgit/pysketcher'
    sys.exit(1)

# Overall dimensions of sketch
H = 15.
W = 17.

drawing_tool.set_coordinate_system(
    xmin=0, xmax=W, ymin=0, ymax=H,
    axis=False)

def sketch(theta, S, mg, drag, t, time_level):
    """
    Draw pendulum sketch with body forces at a time level
    corresponding to time t. The drag force is in
    drag[time_level], the force in the wire is S[time_level],
    the angle is theta[time_level].
    """
    import math
    a = math.degrees(theta[time_level])  # angle in degrees
    L = 0.4*H           # Length of pendulum
    P = (W/2, 0.8*H)  # Fixed rotation point

    mass_pt = path.geometric_features()['end']
    rod = Line(P, mass_pt)

    mass = Circle(center=mass_pt, radius=L/20.)
    mass.set_filled_curves(color='blue')
    rod_vec = rod.geometric_features()['end'] - \
              rod.geometric_features()['start']
    unit_rod_vec = unit_vec(rod_vec)
    mass_symbol = Text('$m$', mass_pt + L/10*unit_rod_vec)
```

```
    rod_start = rod.geometric_features()['start']  # Point P
    vertical = Line(rod_start, rod_start + point(0,-L/3))

    def set_dashed_thin_blackline(*objects):
        """Set linestyle of objects to dashed, black, width=1."""
        for obj in objects:
            obj.set_linestyle('dashed')
            obj.set_linecolor('black')
            obj.set_linewidth(1)

    set_dashed_thin_blackline(vertical)
    set_dashed_thin_blackline(rod)
    angle = Arc_wText(r'$\theta$', rod_start, L/6, -90, a,
                      text_spacing=1/30.)

    magnitude = 1.2*L/2   # length of a unit force in figure
    force = mg[time_level]  # constant (scaled eq: about 1)
    force *= magnitude
    mg_force  = Force(mass_pt, mass_pt + force*point(0,-1),
                      '', text_pos='end')
    force = S[time_level]
    force *= magnitude
    rod_force = Force(mass_pt, mass_pt - force*unit_vec(rod_vec),
                      '', text_pos='end',
                      text_spacing=(0.03, 0.01))
    force = drag[time_level]
    force *= magnitude
    air_force = Force(mass_pt, mass_pt -
                      force*unit_vec((rod_vec[1], -rod_vec[0])),
                      '', text_pos='end',
                      text_spacing=(0.04,0.005))

    body_diagram = Composition(
        {'mg': mg_force, 'S': rod_force, 'air': air_force,
         'rod': rod, 'body': mass
         'vertical': vertical, 'theta': angle,})

    body_diagram.draw(verbose=0)
    drawing_tool.savefig('tmp_%04d.png' % time_level, crop=False)
    # (No cropping: otherwise movies will be very strange!)
```

Making the animated free body diagram It now remains to couple the `simulate` and `sketch` functions. We first run `simulate`:

```
from math import pi, radians, degrees
import numpy as np
alpha = 0.4
period = 2*pi    # Use small theta approximation
T = 12*period    # Simulate for 12 periods
dt = period/40   # 40 time steps per period
a = 70           # Initial amplitude in degrees
Theta = radians(a)

t, theta, omega, S, drag = simulate(alpha, Theta, dt, T)
```

The next step is to run through the time levels in the simulation and make a sketch at each level:

```
for time_level, t_ in enumerate(t):
    sketch(theta, S, mg, drag, t_, time_level)
```

The individual sketches are (by the `sketch` function) saved in files with names `tmp_%04d.png`. These can be combined to videos using (e.g.) `ffmpeg`. A complete function `animate` for running the simulation and creating video files is listed below.

```
def animate():
    # Clean up old plot files
    import os, glob
    for filename in glob.glob('tmp_*.png') + glob.glob('movie.*'):
        os.remove(filename)
    # Solve problem
    from math import pi, radians, degrees
    import numpy as np
    alpha = 0.4
    period = 2*pi   # Use small theta approximation
    T = 12*period   # Simulate for 12 periods
    dt = period/40  # 40 time steps per period
    a = 70          # Initial amplitude in degrees
    Theta = radians(a)

    t, theta, omega, S, drag = simulate(alpha, Theta, dt, T)

    # Visualize drag force 5 times as large
    drag *= 5
    mg = np.ones(S.size)  # Gravity force (needed in sketch)

    # Draw animation
    import time
    for time_level, t_ in enumerate(t):
        sketch(theta, S, mg, drag, t_, time_level)
        time.sleep(0.2)  # Pause between each frame on the screen

    # Make videos
    prog = 'ffmpeg'
    filename = 'tmp_%04d.png'
    fps = 6
    codecs = {'flv': 'flv', 'mp4': 'libx264',
              'webm': 'libvpx', 'ogg': 'libtheora'}
    for ext in codecs:
        lib = codecs[ext]
        cmd = '%(prog)s -i %(filename)s -r %(fps)s ' % vars()
        cmd += '-vcodec %(lib)s movie.%(ext)s' % vars()
        print(cmd)
        os.system(cmd)
```

1.12.7 Motion of an Elastic Pendulum

Consider a pendulum as in Fig. 1.20, but this time the wire is elastic. The length of the wire when it is not stretched is L_0, while $L(t)$ is the stretched length at time t during the motion.

Stretching the elastic wire a distance ΔL gives rise to a spring force $k\Delta L$ in the opposite direction of the stretching. Let $\boldsymbol{n}$ be a unit normal vector along the wire from the point $\boldsymbol{r}_0 = (x_0, y_0)$ and in the direction of $\boldsymbol{i}_\theta$, see Fig. 1.21 for definition of (x_0, y_0) and $\boldsymbol{i}_\theta$. Obviously, we have $\boldsymbol{n} = \boldsymbol{i}_\theta$, but in this modeling of an elastic pendulum we do not need polar coordinates. Instead, it is more straightforward to develop the equation in Cartesian coordinates.

A mathematical expression for $\boldsymbol{n}$ is

$$\boldsymbol{n} = \frac{\boldsymbol{r} - \boldsymbol{r}_0}{L(t)},$$

where $L(t) = ||\boldsymbol{r} - \boldsymbol{r}_0||$ is the current length of the elastic wire. The position vector $\boldsymbol{r}$ in Cartesian coordinates reads $\boldsymbol{r}(t) = x(t)\boldsymbol{i} + y(t)\boldsymbol{j}$, where $\boldsymbol{i}$ and $\boldsymbol{j}$ are unit vectors in the x and y directions, respectively. It is convenient to introduce the Cartesian components n_x and n_y of the normal vector:

$$\boldsymbol{n} = \frac{\boldsymbol{r} - \boldsymbol{r}_0}{L(t)} = \frac{x(t) - x_0}{L(t)}\boldsymbol{i} + \frac{y(t) - y_0}{L(t)}\boldsymbol{j} = n_x\boldsymbol{i} + n_y\boldsymbol{j}\,.$$

The stretch ΔL in the wire is

$$\Delta t = L(t) - L_0\,.$$

The force in the wire is then $-S\boldsymbol{n} = -k\Delta L\boldsymbol{n}$.

The other forces are the gravity and the air resistance, just as in Fig. 1.21. For motion in air we can neglect the added mass and buoyancy effects. The main difference is that we have a *model* for S in terms of the motion (as soon as we have expressed ΔL by $\boldsymbol{r}$). For simplicity, we drop the air resistance term (but Exercise 1.27 asks you to include it).

Newton's second law of motion applied to the body now results in

$$m\ddot{\boldsymbol{r}} = -k(L - L_0)\boldsymbol{n} - mg\boldsymbol{j}\,. \tag{1.132}$$

The two components of (1.132) are

$$\ddot{x} = -\frac{k}{m}(L - L_0)n_x, \tag{1.133}$$

$$\ddot{y} = -\frac{k}{m}(L - L_0)n_y - g\,. \tag{1.134}$$

Remarks about an elastic vs a non-elastic pendulum Note that the derivation of the ODEs for an elastic pendulum is more straightforward than for a classical, non-elastic pendulum, since we avoid the details with polar coordinates, but instead work with Newton's second law directly in Cartesian coordinates. The reason why we can do this is that the elastic pendulum undergoes a general two-dimensional motion where all the forces are known or expressed as functions of $x(t)$ and $y(t)$, such that we get two ordinary differential equations. The motion of the non-elastic pendulum, on the other hand, is constrained: the body has to move along a circular path, and the force S in the wire is unknown.

The non-elastic pendulum therefore leads to a *differential-algebraic* equation, i.e., ODEs for $x(t)$ and $y(t)$ combined with an extra constraint $(x - x_0)^2 + (y - y_0)^2 = L^2$ ensuring that the motion takes place along a circular path. The extra constraint (equation) is compensated by an extra unknown force $-S\boldsymbol{n}$. Differential-algebraic equations are normally hard to solve, especially with pen and paper. Fortunately, for the non-elastic pendulum we can do a trick: in polar coordinates the unknown force S appears only in the radial component of Newton's second law, while the unknown degree of freedom for describing the motion, the angle $\theta(t)$, is completely governed by the asimuthal component. This allows us to decouple the unknowns S and θ. But this is a kind of trick and not a widely applicable method. With an elastic pendulum we use straightforward reasoning with Newton's 2nd law and arrive at a standard ODE problem that (after scaling) is easy to solve on a computer.

Initial conditions What is the initial position of the body? We imagine that first the pendulum hangs in equilibrium in its vertical position, and then it is displaced an angle Θ. The equilibrium position is governed by the ODEs with the accelerations set to zero. The x component leads to $x(t) = x_0$, while the y component gives

$$0 = -\frac{k}{m}(L - L_0)n_y - g = \frac{k}{m}(L(0) - L_0) - g \quad \Rightarrow \quad L(0) = L_0 + mg/k,$$

since $n_y = -11$ in this position. The corresponding y value is then from $n_y = -1$:

$$y(t) = y_0 - L(0) = y_0 - (L_0 + mg/k)\,.$$

Let us now choose (x_0, y_0) such that the body is at the origin in the equilibrium position:

$$x_0 = 0, \quad y_0 = L_0 + mg/k\,.$$

Displacing the body an angle Θ to the right leads to the initial position

$$x(0) = (L_0 + mg/k)\sin\Theta, \quad y(0) = (L_0 + mg/k)(1 - \cos\Theta)\,.$$

The initial velocities can be set to zero: $x'(0) = y'(0) = 0$.

The complete ODE problem We can summarize all the equations as follows:

$$
\begin{aligned}
\ddot{x} &= -\frac{k}{m}(L-L_0)n_x, \\
\ddot{y} &= -\frac{k}{m}(L-L_0)n_y - g, \\
L &= \sqrt{(x-x_0)^2+(y-y_0)^2}, \\
n_x &= \frac{x-x_0}{L}, \\
n_y &= \frac{y-y_0}{L}, \\
x(0) &= (L_0+mg/k)\sin\Theta, \\
x'(0) &= 0, \\
y(0) &= (L_0+mg/k)(1-\cos\Theta), \\
y'(0) &= 0\,.
\end{aligned}
$$

We insert n_x and n_y in the ODEs:

$$\ddot{x} = -\frac{k}{m}\left(1-\frac{L_0}{L}\right)(x-x_0), \tag{1.135}$$

$$\ddot{y} = -\frac{k}{m}\left(1-\frac{L_0}{L}\right)(y-y_0) - g, \tag{1.136}$$

$$L = \sqrt{(x-x_0)^2+(y-y_0)^2}, \tag{1.137}$$

$$x(0) = (L_0+mg/k)\sin\Theta, \tag{1.138}$$

$$x'(0) = 0, \tag{1.139}$$

$$y(0) = (L_0+mg/k)(1-\cos\Theta), \tag{1.140}$$

$$y'(0) = 0\,. \tag{1.141}$$

Scaling The elastic pendulum model can be used to study both an elastic pendulum and a classic, non-elastic pendulum. The latter problem is obtained by letting $k \to \infty$. Unfortunately, a serious problem with the ODEs (1.135)–(1.136) is that for large k, we have a very large factor k/m multiplied by a very small number $1 - L_0/L$, since for large k, $L \approx L_0$ (very small deformations of the wire). The product is subject to significant round-off errors for many relevant physical values of the parameters. To circumvent the problem, we introduce a scaling. This will also remove physical parameters from the problem such that we end up with only one dimensionless parameter, closely related to the elasticity of the wire. Simulations can then be done by setting just this dimensionless parameter.

The characteristic length can be taken such that in equilibrium, the scaled length is unity, i.e., the characteristic length is $L_0 + mg/k$:

$$\bar{x} = \frac{x}{L_0+mg/k}, \quad \bar{y} = \frac{y}{L_0+mg/k}\,.$$

We must then also work with the scaled length $\bar{L} = L/(L_0 + mg/k)$.

Introducing $\bar{t} = t/t_c$, where t_c is a characteristic time we have to decide upon later, one gets

$$\begin{aligned}
\frac{d^2\bar{x}}{d\bar{t}^2} &= -t_c^2\frac{k}{m}\left(1 - \frac{L_0}{L_0 + mg/k}\frac{1}{\bar{L}}\right)\bar{x},\\
\frac{d^2\bar{y}}{d\bar{t}^2} &= -t_c^2\frac{k}{m}\left(1 - \frac{L_0}{L_0 + mg/k}\frac{1}{\bar{L}}\right)(\bar{y} - 1) - t_c^2\frac{g}{L_0 + mg/k},\\
\bar{L} &= \sqrt{\bar{x}^2 + (\bar{y} - 1)^2},\\
\bar{x}(0) &= \sin\Theta,\\
\bar{x}'(0) &= 0,\\
\bar{y}(0) &= 1 - \cos\Theta,\\
\bar{y}'(0) &= 0\,.
\end{aligned}$$

For a non-elastic pendulum with small angles, we know that the frequency of the oscillations are $\omega = \sqrt{L/g}$. It is therefore natural to choose a similar expression here, either the length in the equilibrium position,

$$t_c^2 = \frac{L_0 + mg/k}{g}\,.$$

or simply the unstretched length,

$$t_c^2 = \frac{L_0}{g}\,.$$

These quantities are not very different (since the elastic model is valid only for quite small elongations), so we take the latter as it is the simplest one.

The ODEs become

$$\begin{aligned}
\frac{d^2\bar{x}}{d\bar{t}^2} &= -\frac{L_0 k}{mg}\left(1 - \frac{L_0}{L_0 + mg/k}\frac{1}{\bar{L}}\right)\bar{x},\\
\frac{d^2\bar{y}}{d\bar{t}^2} &= -\frac{L_0 k}{mg}\left(1 - \frac{L_0}{L_0 + mg/k}\frac{1}{\bar{L}}\right)(\bar{y} - 1) - \frac{L_0}{L_0 + mg/k},\\
\bar{L} &= \sqrt{\bar{x}^2 + (\bar{y} - 1)^2}\,.
\end{aligned}$$

We can now identify a dimensionless number

$$\beta = \frac{L_0}{L_0 + mg/k} = \frac{1}{1 + \frac{mg}{L_0 k}},$$

which is the ratio of the unstretched length and the stretched length in equilibrium. The non-elastic pendulum will have $\beta = 1$ ($k \to \infty$). With β the ODEs read

$$\frac{d^2\bar{x}}{d\bar{t}^2} = -\frac{\beta}{1-\beta}\left(1-\frac{\beta}{\bar{L}}\right)\bar{x}, \quad (1.142)$$

$$\frac{d^2\bar{y}}{d\bar{t}^2} = -\frac{\beta}{1-\beta}\left(1-\frac{\beta}{\bar{L}}\right)(\bar{y}-1)-\beta, \quad (1.143)$$

$$\bar{L} = \sqrt{\bar{x}^2+(\bar{y}-1)^2}, \quad (1.144)$$

$$\bar{x}(0) = (1+\epsilon)\sin\Theta, \quad (1.145)$$

$$\frac{d\bar{x}}{d\bar{t}}(0) = 0, \quad (1.146)$$

$$\bar{y}(0) = 1-(1+\epsilon)\cos\Theta, \quad (1.147)$$

$$\frac{d\bar{y}}{d\bar{t}}(0) = 0, \quad (1.148)$$

We have here added a parameter ϵ, which is an additional downward stretch of the wire at $t = 0$. This parameter makes it possible to do a desired test: vertical oscillations of the pendulum. Without ϵ, starting the motion from $(0, 0)$ with zero velocity will result in $x = y = 0$ for all times (also a good test!), but with an initial stretch so the body's position is $(0, \epsilon)$, we will have oscillatory vertical motion with amplitude ϵ (see Exercise 1.26).

Remark on the non-elastic limit We immediately see that as $k \to \infty$ (i.e., we obtain a non-elastic pendulum), $\beta \to 1$, $\bar{L} \to 1$, and we have very small values $1 - \beta\bar{L}^{-1}$ divided by very small values $1 - \beta$ in the ODEs. However, it turns out that we can set β very close to one and obtain a path of the body that within the visual accuracy of a plot does not show any elastic oscillations. (Should the division of very small values become a problem, one can study the limit by L'Hospital's rule:

$$\lim_{\beta\to 1}\frac{1-\beta\bar{L}^{-1}}{1-\beta} = \frac{1}{\bar{L}},$$

and use the limit $\bar{L}^{-1}$ in the ODEs for β values very close to 1.)

1.12.8 Vehicle on a Bumpy Road

We consider a very simplistic vehicle, on one wheel, rolling along a bumpy road. The oscillatory nature of the road will induce an external forcing on the spring system in the vehicle and cause vibrations. Figure 1.22 outlines the situation.

To derive the equation that governs the motion, we must first establish the position vector of the black mass at the top of the spring. Suppose the spring has length L without any elongation or compression, suppose the radius of the wheel is R, and suppose the height of the black mass at the top is H. With the aid of the $\boldsymbol{r}_0$ vector in Fig. 1.22, the position $\boldsymbol{r}$ of the center point of the mass is

$$\boldsymbol{r} = \boldsymbol{r}_0 + 2R\boldsymbol{j} + L\boldsymbol{j} + u\boldsymbol{j} + \frac{1}{2}H\boldsymbol{j}, \quad (1.149)$$

Fig. 1.22 Sketch of one-wheel vehicle on a bumpy road

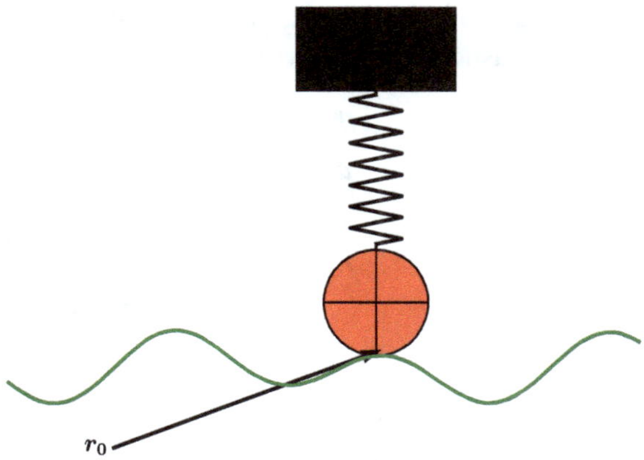

where u is the elongation or compression in the spring according to the (unknown and to be computed) vertical displacement u relative to the road. If the vehicle travels with constant horizontal velocity v and $h(x)$ is the shape of the road, then the vector $\boldsymbol{r}_0$ is

$$\boldsymbol{r}_0 = vt\boldsymbol{i} + h(vt)\boldsymbol{j},$$

if the motion starts from $x = 0$ at time $t = 0$.

The forces on the mass is the gravity, the spring force, and an optional damping force that is proportional to the vertical velocity $\dot{u}$. Newton's second law of motion then tells that

$$m\ddot{\boldsymbol{r}} = -mg\boldsymbol{j} - s(u) - b\dot{u}\boldsymbol{j}\,.$$

This leads to

$$m\ddot{u} = -s(u) - b\dot{u} - mg - mh''(vt)v^2\,.$$

To simplify a little bit, we omit the gravity force mg in comparison with the other terms. Introducing u' for $\dot{u}$ then gives a standard damped, vibration equation with external forcing:

$$mu'' + bu' + s(u) = -mh''(vt)v^2\,. \tag{1.150}$$

Since the road is normally known just as a set of array values, h'' must be computed by finite differences. Let Δx be the spacing between measured values $h_i = h(i\Delta x)$ on the road. The discrete second-order derivative h'' reads

$$q_i = \frac{h_{i-1} - 2h_i + h_{i+1}}{\Delta x^2}, \quad i = 1, \ldots, N_x - 1\,.$$

We may for maximum simplicity set the end points as $q_0 = q_1$ and $q_{N_x} = q_{N_x-1}$. The term $-mh''(vt)v^2$ corresponds to a force with discrete time values

$$F^n = -mq_n v^2, \quad \Delta t = v^{-1}\Delta x\,.$$

This force can be directly used in a numerical model

$$[mD_tD_tu + bD_{2t}u + s(u) = F]^n\,.$$

Software for computing u and also making an animated sketch of the motion like we did in Sect. 1.12.6 is found in a separate project on the web: `https://github.com/hplgit/bumpy`. You may start looking at the tutorial[20].

1.12.9 Bouncing Ball

A bouncing ball is a ball in free vertically fall until it impacts the ground, but during the impact, some kinetic energy is lost, and a new motion upwards with reduced velocity starts. After the motion is retarded, a new free fall starts, and the process is repeated. At some point the velocity close to the ground is so small that the ball is considered to be finally at rest.

The motion of the ball falling in air is governed by Newton's second law $F = ma$, where a is the acceleration of the body, m is the mass, and F is the sum of all forces. Here, we neglect the air resistance so that gravity $-mg$ is the only force. The height of the ball is denoted by h and v is the velocity. The relations between h, v, and a,

$$h'(t) = v(t), \quad v'(t) = a(t),$$

combined with Newton's second law gives the ODE model

$$h''(t) = -g, \tag{1.151}$$

or expressed alternatively as a system of first-order equations:

$$v'(t) = -g, \tag{1.152}$$

$$h'(t) = v(t)\,. \tag{1.153}$$

These equations govern the motion as long as the ball is away from the ground by a small distance $\epsilon_h > 0$. When $h < \epsilon_h$, we have two cases.

1. The ball impacts the ground, recognized by a sufficiently large negative velocity ($v < -\epsilon_v$). The velocity then changes sign and is reduced by a factor C_R, known as the coefficient of restitution[21]. For plotting purposes, one may set $h = 0$.
2. The motion stops, recognized by a sufficiently small velocity ($|v| < \epsilon_v$) close to the ground.

1.12.10 Two-Body Gravitational Problem

Consider two astronomical objects A and B that attract each other by gravitational forces. A and B could be two stars in a binary system, a planet orbiting a star, or a moon orbiting a planet. Each object is acted upon by the gravitational force due to the other object. Consider motion in a plane (for simplicity) and let (x_A, y_A) and (x_B, y_B) be the positions of object A and B, respectively.

---

[20] http://hplgit.github.io/bumpy/doc/pub/bumpy.pdf

[21] http://en.wikipedia.org/wiki/Coefficient_of_restitution

The governing equations Newton's second law of motion applied to each object is all we need to set up a mathematical model for this physical problem:

$$m_A\ddot{\boldsymbol{x}}_A = \boldsymbol{F}, \quad (1.154)$$

$$m_B\ddot{\boldsymbol{x}}_B = -\boldsymbol{F}, \quad (1.155)$$

where F is the gravitational force

$$\boldsymbol{F} = \frac{Gm_Am_B}{||\boldsymbol{r}||^3}\boldsymbol{r},$$

where

$$\boldsymbol{r}(t) = \boldsymbol{x}_B(t) - \boldsymbol{x}_A(t),$$

and G is the gravitational constant: $G = 6.674 \cdot 10^{-11}\,\mathrm{Nm}^2/\mathrm{kg}^2$.

Scaling A problem with these equations is that the parameters are very large (m_A, m_B, $||\boldsymbol{r}||$) or very small (G). The rotation time for binary stars can be very small and large as well. It is therefore advantageous to scale the equations. A natural length scale could be the initial distance between the objects: $L = \boldsymbol{r}(0)$. We write the dimensionless quantities as

$$\bar{\boldsymbol{x}}_A = \frac{\boldsymbol{x}_A}{L}, \quad \bar{\boldsymbol{x}}_B = \frac{\boldsymbol{x}_B}{L}, \quad \bar{t} = \frac{t}{t_c}.$$

The gravity force is transformed to

$$\boldsymbol{F} = \frac{Gm_Am_B}{L^2||\bar{\boldsymbol{r}}||^3}\bar{\boldsymbol{r}}, \quad \bar{\boldsymbol{r}} = \bar{\boldsymbol{x}}_B - \bar{\boldsymbol{x}}_A,$$

so the first ODE for $\boldsymbol{x}_A$ becomes

$$\frac{d^2\bar{\boldsymbol{x}}_A}{d\bar{t}^2} = \frac{Gm_Bt_c^2}{L^3}\frac{\bar{\boldsymbol{r}}}{||\bar{\boldsymbol{r}}||^3}.$$

Assuming that quantities with a bar and their derivatives are around unity in size, it is natural to choose t_c such that the fraction $Gm_Bt_c/L^2 = 1$:

$$t_c = \sqrt{\frac{L^3}{Gm_B}}.$$

From the other equation for $\boldsymbol{x}_B$ we get another candidate for t_c with m_A instead of m_B. Which mass we choose play a role if $m_A \ll m_B$ or $m_B \ll m_A$. One solution is to use the sum of the masses:

$$t_c = \sqrt{\frac{L^3}{G(m_A+m_B)}}.$$

Taking a look at Kepler's laws[22] of planetary motion, the orbital period for a planet around the star is given by the t_c above, except for a missing factor of 2π, but that means that t_c^{-1} is just the angular frequency of the motion. Our characteristic time t_c is therefore highly relevant. Introducing the dimensionless number

$$\alpha = \frac{m_A}{m_B},$$

we can write the dimensionless ODE as

$$\frac{d^2\bar{\boldsymbol{x}}_A}{d\bar{t}^2} = \frac{1}{1+\alpha}\frac{\bar{\boldsymbol{r}}}{||\bar{\boldsymbol{r}}||^3}, \tag{1.156}$$

$$\frac{d^2\bar{\boldsymbol{x}}_B}{d\bar{t}^2} = \frac{1}{1+\alpha^{-1}}\frac{\bar{\boldsymbol{r}}}{||\bar{\boldsymbol{r}}||^3}. \tag{1.157}$$

In the limit $m_A \ll m_B$, i.e., $\alpha \ll 1$, object B stands still, say $\bar{\boldsymbol{x}}_B = 0$, and object A orbits according to

$$\frac{d^2\bar{\boldsymbol{x}}_A}{d\bar{t}^2} = -\frac{\bar{\boldsymbol{x}}_A}{||\bar{\boldsymbol{x}}_A||^3}.$$

Solution in a special case: planet orbiting a star To better see the motion, and that our scaling is reasonable, we introduce polar coordinates r and θ:

$$\bar{\boldsymbol{x}}_A = r\cos\theta\boldsymbol{i} + r\sin\theta\boldsymbol{j},$$

which means $\bar{\boldsymbol{x}}_A$ can be written as $\bar{\boldsymbol{x}}_A = r\boldsymbol{i}_r$. Since

$$\frac{d}{dt}\boldsymbol{i}_r = \dot{\theta}\boldsymbol{i}_\theta, \quad \frac{d}{dt}\boldsymbol{i}_\theta = -\dot{\theta}\boldsymbol{i}_r,$$

we have

$$\frac{d^2\bar{\boldsymbol{x}}_A}{d\bar{t}^2} = (\ddot{r} - r\dot{\theta}^2)\boldsymbol{i}_r + (r\ddot{\theta} + 2\dot{r}\dot{\theta})\boldsymbol{i}_\theta.$$

The equation of motion for mass A is then

$$\ddot{r} - r\dot{\theta}^2 = -\frac{1}{r^2},$$

$$r\ddot{\theta} + 2\dot{r}\dot{\theta} = 0.$$

The special case of circular motion, $r = 1$, fulfills the equations, since the latter equation then gives $\dot{\theta} = \text{const}$ and the former then gives $\dot{\theta} = 1$, i.e., the motion is $r(t) = 1$, $\theta(t) = t$, with unit angular frequency as expected and period 2π as expected.

[22] https://en.wikipedia.org/wiki/Kepler%27s_laws_of_planetary_motion

1.12.11 Electric Circuits

Although the term "mechanical vibrations" is used in the present book, we must mention that the same type of equations arise when modeling electric circuits. The current $I(t)$ in a circuit with an inductor with inductance L, a capacitor with capacitance C, and overall resistance R, is governed by

$$\ddot{I} + \frac{R}{L}\dot{I} + \frac{1}{LC}I = \dot{V}(t), \tag{1.158}$$

where $V(t)$ is the voltage source powering the circuit. This equation has the same form as the general model considered in Sect. 1.10 if we set $u = I$, $f(u') = bu'$ and define $b = R/L$, $s(u) = L^{-1}C^{-1}u$, and $F(t) = \dot{V}(t)$.

1.13 Exercises

Exercise 1.22: Simulate resonance
We consider the scaled ODE model (1.122) from Sect. 1.12.2. After scaling, the amplitude of u will have a size about unity as time grows and the effect of the initial conditions die out due to damping. However, as $\gamma \to 1$, the amplitude of u increases, especially if β is small. This effect is called *resonance*. The purpose of this exercise is to explore resonance.

a) Figure out how the `solver` function in `vib.py` can be called for the scaled ODE (1.122).
b) Run $\gamma = 5, 1.5, 1.1, 1$ for $\beta = 0.005, 0.05, 0.2$. For each β value, present an image with plots of $u(t)$ for the four γ values.

Filename: `resonance`.

Exercise 1.23: Simulate oscillations of a sliding box
Consider a sliding box on a flat surface as modeled in Sect. 1.12.3. As spring force we choose the nonlinear formula

$$s(u) = \frac{k}{\alpha}\tanh(\alpha u) = ku + \frac{1}{3}\alpha^2 ku^3 + \frac{2}{15}\alpha^4 ku^5 + \mathcal{O}(u^6).$$

a) Plot $g(u) = \alpha^{-1}\tanh(\alpha u)$ for various values of α. Assume $u \in [-1, 1]$.
b) Scale the equations using I as scale for u and $\sqrt{m/k}$ as time scale.
c) Implement the scaled model in b). Run it for some values of the dimensionless parameters.

Filename: `sliding_box`.

Exercise 1.24: Simulate a bouncing ball
Section 1.12.9 presents a model for a bouncing ball. Choose one of the two ODE formulation, (1.151) or (1.152)–(1.153), and simulate the motion of a bouncing ball. Plot $h(t)$. Think about how to plot $v(t)$.

Hint A naive implementation may get stuck in repeated impacts for large time step sizes. To avoid this situation, one can introduce a state variable that holds the mode of the motion: free fall, impact, or rest. Two consecutive impacts imply that the motion has stopped.
Filename: `bouncing_ball`.

Exercise 1.25: Simulate a simple pendulum
Simulation of simple pendulum can be carried out by using the mathematical model derived in Sect. 1.12.5 and calling up functionality in the `vib.py` file (i.e., solve the second-order ODE by centered finite differences).

a) Scale the model. Set up the dimensionless governing equation for θ and expressions for dimensionless drag and wire forces.
b) Write a function for computing θ and the dimensionless drag force and the force in the wire, using the `solver` function in the `vib.py` file. Plot these three quantities below each other (in subplots) so the graphs can be compared. Run two cases, first one in the limit of Θ small and no drag, and then a second one with $\Theta = 40$ degrees and $\alpha = 0.8$.

Filename: `simple_pendulum`.

Exercise 1.26: Simulate an elastic pendulum
Section 1.12.7 describes a model for an elastic pendulum, resulting in a system of two ODEs. The purpose of this exercise is to implement the scaled model, test the software, and generalize the model.

a) Write a function `simulate` that can simulate an elastic pendulum using the scaled model. The function should have the following arguments:

```
def simulate(
    beta=0.9,                 # dimensionless parameter
    Theta=30,                 # initial angle in degrees
    epsilon=0,                # initial stretch of wire
    num_periods=6,            # simulate for num_periods
    time_steps_per_period=60, # time step resolution
    plot=True,                # make plots or not
    ):
```

To set the total simulation time and the time step, we use our knowledge of the scaled, classical, non-elastic pendulum: $u'' + u = 0$, with solution $u = \Theta \cos \bar{t}$. The period of these oscillations is $P = 2\pi$ and the frequency is unity. The time for simulation is taken as `num_periods` times P. The time step is set as P divided by `time_steps_per_period`.
The `simulate` function should return the arrays of x, y, θ, and t, where $\theta = \tan^{-1}(x/(1-y))$ is the angular displacement of the elastic pendulum corresponding to the position (x, y).
If `plot` is `True`, make a plot of $\bar{y}(\bar{t})$ versus $\bar{x}(\bar{t})$, i.e., the physical motion of the mass at $(\bar{x}, \bar{y})$. Use the equal aspect ratio on the axis such that we get a physically correct picture of the motion. Also make a plot of $\theta(\bar{t})$, where θ is

measured in degrees. If $\Theta < 10$ degrees, add a plot that compares the solutions of the scaled, classical, non-elastic pendulum and the elastic pendulum ($\theta(t)$). Although the mathematics here employs a bar over scaled quantities, the code should feature plain names `x` for $\bar{x}$, `y` for $\bar{y}$, and `t` for $\bar{t}$ (rather than `x_bar`, etc.). These variable names make the code easier to read and compare with the mathematics.

Hint 1 Equal aspect ratio is set by `plt.gca().set_aspect('equal')` in Matplotlib (`import matplotlib.pyplot as plt`) and in SciTools by the command `plt.plot(..., daspect=[1,1,1], daspectmode='equal')` (provided you have done `import scitools.std as plt`).

Hint 2 If you want to use Odespy to solve the equations, order the ODEs like $\dot{\bar{x}}, \bar{x}, \dot{\bar{y}}, \bar{y}$ such that `odespy.EulerCromer` can be applied.

b) Write a test function for testing that $\Theta = 0$ and $\epsilon = 0$ gives $x = y = 0$ for all times.
c) Write another test function for checking that the pure vertical motion of the elastic pendulum is correct. Start with simplifying the ODEs for pure vertical motion and show that $\bar{y}(\bar{t})$ fulfills a vibration equation with frequency $\sqrt{\beta/(1-\beta)}$. Set up the exact solution.
   Write a test function that uses this special case to verify the `simulate` function. There will be numerical approximation errors present in the results from `simulate` so you have to believe in correct results and set a (low) tolerance that corresponds to the computed maximum error. Use a small Δt to obtain a small numerical approximation error.
d) Make a function `demo(beta, Theta)` for simulating an elastic pendulum with a given β parameter and initial angle Θ. Use 600 time steps per period to get every accurate results, and simulate for 3 periods.

Filename: `elastic_pendulum`.

Exercise 1.27: Simulate an elastic pendulum with air resistance

This is a continuation Exercise 1.26. Air resistance on the body with mass m can be modeled by the force $-\frac{1}{2}\varrho C_D A|\boldsymbol{v}|\boldsymbol{v}$, where C_D is a drag coefficient (0.2 for a sphere), ϱ is the density of air ($1.2\,\text{kg}\,\text{m}^{-3}$), A is the cross section area ($A = \pi R^2$ for a sphere, where R is the radius), and $\boldsymbol{v}$ is the velocity of the body. Include air resistance in the original model, scale the model, write a function `simulate_drag` that is a copy of the `simulate` function from Exercise 1.26, but with the new ODEs included, and show plots of how air resistance influences the motion.
Filename: `elastic_pendulum_drag`.

Remarks Test functions are challenging to construct for the problem with air resistance. You can reuse the tests from Exercise 1.27 for `simulate_drag`, but these tests does not verify the new terms arising from air resistance.

Exercise 1.28: Implement the PEFRL algorithm
We consider the motion of a planet around a star (Sect. 1.12.10). The simplified case where one mass is very much bigger than the other and one object is at rest, results in the scaled ODE model

$$\ddot{x} + (x^2 + y^2)^{-3/2}x = 0,$$

$$\ddot{y} + (x^2 + y^2)^{-3/2}y = 0\,.$$

a) It is easy to show that $x(t)$ and $y(t)$ go like sine and cosine functions. Use this idea to derive the exact solution.
b) One believes that a planet may orbit a star for billions of years. We are now interested in how accurate methods we actually need for such calculations. A first task is to determine what the time interval of interest is in scaled units. Take the earth and sun as typical objects and find the characteristic time used in the scaling of the equations ($t_c = \sqrt{L^3/(mG)}$), where m is the mass of the sun, L is the distance between the sun and the earth, and G is the gravitational constant. Find the scaled time interval corresponding to one billion years.
c) Solve the equations using 4th-order Runge-Kutta and the Euler-Cromer methods. You may benefit from applying Odespy for this purpose. With each solver, simulate 10,000 orbits and print the maximum position error and CPU time as a function of time step. Note that the maximum position error does not necessarily occur at the end of the simulation. The position error achieved with each solver will depend heavily on the size of the time step. Let the time step correspond to 200, 400, 800 and 1600 steps per orbit, respectively. Are the results as expected? Explain briefly. When you develop your program, have in mind that it will be extended with an implementation of the other algorithms (as requested in d) and e) later) and experiments with this algorithm as well.
d) Implement a solver based on the PEFRL method from Sect. 1.10.11. Verify its 4th-order convergence using an equation $u'' + u = 0$.
e) The simulations done previously with the 4th-order Runge-Kutta and Euler-Cromer are now to be repeated with the PEFRL solver, so the code must be extended accordingly. Then run the simulations and comment on the performance of PEFRL compared to the other two.
f) Use the PEFRL solver to simulate 100,000 orbits with a fixed time step corresponding to 1600 steps per period. Record the maximum error within each subsequent group of 1000 orbits. Plot these errors and fit (least squares) a mathematical function to the data. Print also the total CPU time spent for all 100,000 orbits.
   Now, predict the error and required CPU time for a simulation of 1 billion years (orbits). Is it feasible on today's computers to simulate the planetary motion for one billion years?

Filename: `vib_PEFRL`.

Remarks This exercise investigates whether it is feasible to predict planetary motion for the life time of a solar system.

2

A Complete Study of Wave Equations

A very wide range of physical processes lead to wave motion, where signals are propagated through a medium in space and time, normally with little or no permanent movement of the medium itself. The shape of the signals may undergo changes as they travel through matter, but usually not so much that the signals cannot be recognized at some later point in space and time. Many types of wave motion can be described by the equation $u_{tt} = \nabla \cdot (c^2 \nabla u) + f$, which we will solve in the forthcoming text by finite difference methods.

2.1 Simulation of Waves on a String

We begin our study of wave equations by simulating one-dimensional waves on a string, say on a guitar or violin. Let the string in the undeformed state coincide with the interval $[0, L]$ on the x axis, and let $u(x,t)$ be the displacement at time t in the y direction of a point initially at x. The displacement function u is governed by the mathematical model

$$\frac{\partial^2 u}{\partial t^2} = c^2 \frac{\partial^2 u}{\partial x^2}, \qquad x \in (0, L),\ t \in (0, T] \tag{2.1}$$

$$u(x, 0) = I(x), \qquad x \in [0, L] \tag{2.2}$$

$$\frac{\partial}{\partial t} u(x, 0) = 0, \qquad x \in [0, L] \tag{2.3}$$

$$u(0, t) = 0, \qquad t \in (0, T] \tag{2.4}$$

$$u(L, t) = 0, \qquad t \in (0, T]\,. \tag{2.5}$$

The constant c and the function $I(x)$ must be prescribed.

Equation (2.1) is known as the one-dimensional *wave equation*. Since this PDE contains a second-order derivative in time, we need *two initial conditions*. The condition (2.2) specifies the initial shape of the string, $I(x)$, and (2.3) expresses that the initial velocity of the string is zero. In addition, PDEs need *boundary conditions*, given here as (2.4) and (2.5). These two conditions specify that the string is fixed at the ends, i.e., that the displacement u is zero.

The solution $u(x,t)$ varies in space and time and describes waves that move with velocity c to the left and right.

Sometimes we will use a more compact notation for the partial derivatives to save space:

$$u_t = \frac{\partial u}{\partial t}, \quad u_{tt} = \frac{\partial^2 u}{\partial t^2}, \tag{2.6}$$

and similar expressions for derivatives with respect to other variables. Then the wave equation can be written compactly as $u_{tt} = c^2 u_{xx}$.

The PDE problem (2.1)–(2.5) will now be discretized in space and time by a finite difference method.

2.1.1 Discretizing the Domain

The temporal domain $[0, T]$ is represented by a finite number of mesh points

$$0 = t_0 < t_1 < t_2 < \cdots < t_{N_t-1} < t_{N_t} = T\,. \tag{2.7}$$

Similarly, the spatial domain $[0, L]$ is replaced by a set of mesh points

$$0 = x_0 < x_1 < x_2 < \cdots < x_{N_x-1} < x_{N_x} = L\,. \tag{2.8}$$

One may view the mesh as two-dimensional in the x, t plane, consisting of points (x_i, t_n), with $i = 0, \ldots, N_x$ and $n = 0, \ldots, N_t$.

Uniform meshes For uniformly distributed mesh points we can introduce the constant mesh spacings Δt and Δx. We have that

$$x_i = i\Delta x,\ i = 0, \ldots, N_x, \quad t_n = n\Delta t,\ n = 0, \ldots, N_t\,. \tag{2.9}$$

We also have that $\Delta x = x_i - x_{i-1}$, $i = 1, \ldots, N_x$, and $\Delta t = t_n - t_{n-1}$, $n = 1, \ldots, N_t$. Figure 2.1 displays a mesh in the x, t plane with $N_t = 5$, $N_x = 5$, and constant mesh spacings.

2.1.2 The Discrete Solution

The solution $u(x, t)$ is sought at the mesh points. We introduce the mesh function u_i^n, which approximates the exact solution at the mesh point (x_i, t_n) for $i = 0, \ldots, N_x$ and $n = 0, \ldots, N_t$. Using the finite difference method, we shall develop algebraic equations for computing the mesh function.

2.1.3 Fulfilling the Equation at the Mesh Points

In the finite difference method, we relax the condition that (2.1) holds at all points in the space-time domain $(0, L) \times (0, T]$ to the requirement that the PDE is fulfilled at the *interior* mesh points only:

$$\frac{\partial^2}{\partial t^2} u(x_i, t_n) = c^2 \frac{\partial^2}{\partial x^2} u(x_i, t_n), \tag{2.10}$$

for $i = 1,\ldots,N_x - 1$ and $n = 1,\ldots,N_t - 1$. For $n = 0$ we have the initial conditions $u = I(x)$ and $u_t = 0$, and at the boundaries $i = 0, N_x$ we have the boundary condition $u = 0$.

2.1.4 Replacing Derivatives by Finite Differences

The second-order derivatives can be replaced by central differences. The most widely used difference approximation of the second-order derivative is

$$\frac{\partial^2}{\partial t^2}u(x_i,t_n) \approx \frac{u_i^{n+1} - 2u_i^n + u_i^{n-1}}{\Delta t^2}\,.$$

It is convenient to introduce the finite difference operator notation

$$[D_tD_tu]_i^n = \frac{u_i^{n+1} - 2u_i^n + u_i^{n-1}}{\Delta t^2}\,.$$

A similar approximation of the second-order derivative in the x direction reads

$$\frac{\partial^2}{\partial x^2}u(x_i,t_n) \approx \frac{u_{i+1}^n - 2u_i^n + u_{i-1}^n}{\Delta x^2} = [D_xD_xu]_i^n\,.$$

Algebraic version of the PDE We can now replace the derivatives in (2.10) and get

$$\frac{u_i^{n+1} - 2u_i^n + u_i^{n-1}}{\Delta t^2} = c^2\frac{u_{i+1}^n - 2u_i^n + u_{i-1}^n}{\Delta x^2}, \tag{2.11}$$

or written more compactly using the operator notation:

$$[D_tD_tu = c^2D_xD_x]_i^n\,. \tag{2.12}$$

Interpretation of the equation as a stencil A characteristic feature of (2.11) is that it involves u values from neighboring points only: u_i^{n+1}, $u_{i\pm1}^n$, u_i^n, and u_i^{n-1}. The circles in Fig. 2.1 illustrate such neighboring mesh points that contribute to an algebraic equation. In this particular case, we have sampled the PDE at the point $(2,2)$ and constructed (2.11), which then involves a coupling of u_1^2, u_2^3, u_2^2, u_2^1, and u_3^2. The term *stencil* is often used about the algebraic equation at a mesh point, and the geometry of a typical stencil is illustrated in Fig. 2.1. One also often refers to the algebraic equations as *discrete equations*, *(finite) difference equations* or a *finite difference scheme*.

Algebraic version of the initial conditions We also need to replace the derivative in the initial condition (2.3) by a finite difference approximation. A centered difference of the type

$$\frac{\partial}{\partial t}u(x_i,t_0) \approx \frac{u_i^1 - u_i^{-1}}{2\Delta t} = [D_{2t}u]_i^0,$$

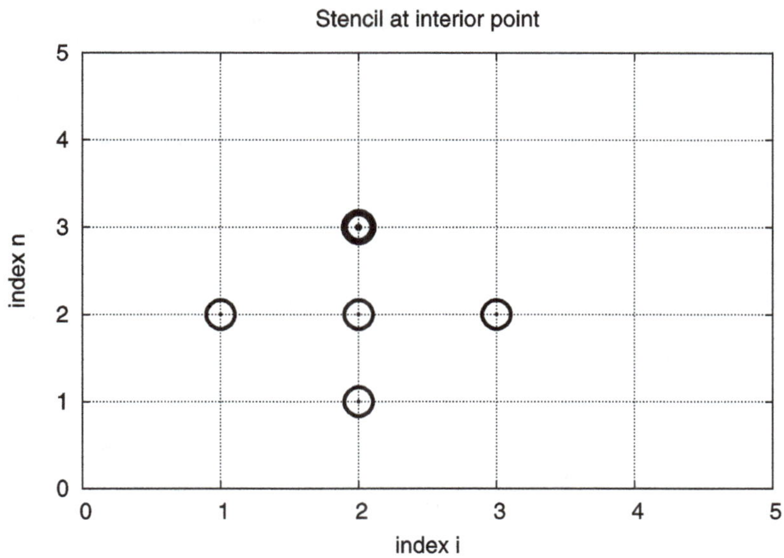

Fig. 2.1 Mesh in space and time. The circles show points connected in a finite difference equation

seems appropriate. Writing out this equation and ordering the terms give

$$u_i^{-1} = u_i^1, \quad i = 0, \ldots, N_x\,. \tag{2.13}$$

The other initial condition can be computed by

$$u_i^0 = I(x_i), \quad i = 0, \ldots, N_x\,.$$

2.1.5 Formulating a Recursive Algorithm

We assume that u_i^n and u_i^{n-1} are available for $i = 0, \ldots, N_x$. The only unknown quantity in (2.11) is therefore u_i^{n+1}, which we now can solve for:

$$u_i^{n+1} = -u_i^{n-1} + 2u_i^n + C^2\left(u_{i+1}^n - 2u_i^n + u_{i-1}^n\right)\,. \tag{2.14}$$

We have here introduced the parameter

$$C = c\frac{\Delta t}{\Delta x}, \tag{2.15}$$

known as the *Courant number*.

C is the key parameter in the discrete wave equation

We see that the discrete version of the PDE features only one parameter, C, which is therefore the key parameter, together with N_x, that governs the quality of the numerical solution (see Sect. 2.10 for details). Both the primary physical parameter c and the numerical parameters Δx and Δt are lumped together in C. Note that C is a dimensionless parameter.

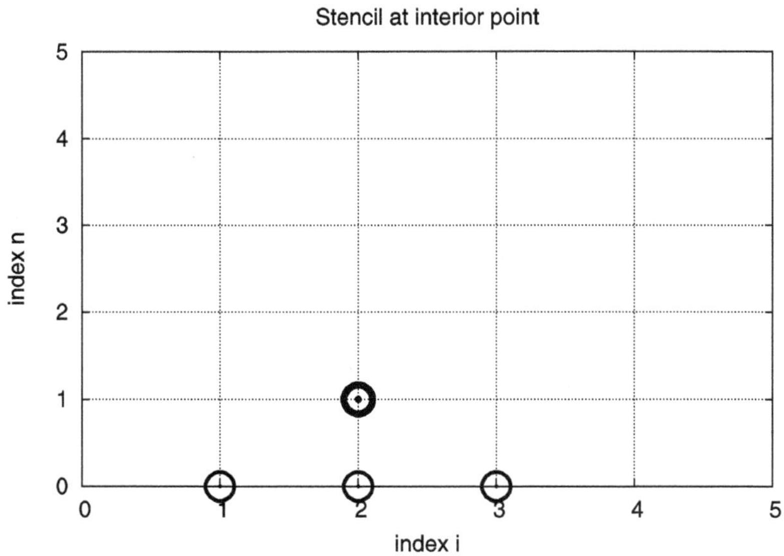

Fig. 2.2 Modified stencil for the first time step

Given that u_i^{n-1} and u_i^n are known for $i = 0, \ldots, N_x$, we find new values at the next time level by applying the formula (2.14) for $i = 1, \ldots, N_x - 1$. Figure 2.1 illustrates the points that are used to compute u_2^3. For the boundary points, $i = 0$ and $i = N_x$, we apply the boundary conditions $u_i^{n+1} = 0$.

Even though sound reasoning leads up to (2.14), there is still a minor challenge with it that needs to be resolved. Think of the very first computational step to be made. The scheme (2.14) is supposed to start at $n = 1$, which means that we compute u^2 from u^1 and u^0. Unfortunately, we do not know the value of u^1, so how to proceed? A standard procedure in such cases is to apply (2.14) also for $n = 0$. This immediately seems strange, since it involves u_i^{-1}, which is an undefined quantity outside the time mesh (and the time domain). However, we can use the initial condition (2.13) in combination with (2.14) when $n = 0$ to eliminate u_i^{-1} and arrive at a special formula for u_i^1:

$$u_i^1 = u_i^0 - \frac{1}{2}C^2\left(u_{i+1}^0 - 2u_i^0 + u_{i-1}^0\right). \tag{2.16}$$

Figure 2.2 illustrates how (2.16) connects four instead of five points: u_2^1, u_1^0, u_2^0, and u_3^0.

We can now summarize the computational algorithm:

1. Compute $u_i^0 = I(x_i)$ for $i = 0, \ldots, N_x$
2. Compute u_i^1 by (2.16) for $i = 1, 2, \ldots, N_x - 1$ and set $u_i^1 = 0$ for the boundary points given by $i = 0$ and $i = N_x$,
3. For each time level $n = 1, 2, \ldots, N_t - 1$
   (a) apply (2.14) to find u_i^{n+1} for $i = 1, \ldots, N_x - 1$
   (b) set $u_i^{n+1} = 0$ for the boundary points having $i = 0$, $i = N_x$.

The algorithm essentially consists of moving a finite difference stencil through all the mesh points, which can be seen as an animation in a web page[1] or a movie file[2].

2.1.6 Sketch of an Implementation

The algorithm only involves the three most recent time levels, so we need only three arrays for u_i^{n+1}, u_i^n, and u_i^{n-1}, $i = 0, \ldots, N_x$. Storing all the solutions in a two-dimensional array of size $(N_x + 1) \times (N_t + 1)$ would be possible in this simple one-dimensional PDE problem, but is normally out of the question in three-dimensional (3D) and large two-dimensional (2D) problems. We shall therefore, in all our PDE solving programs, have the unknown in memory at as few time levels as possible.

In a Python implementation of this algorithm, we use the array elements `u[i]` to store u_i^{n+1}, `u_n[i]` to store u_i^n, and `u_nm1[i]` to store u_i^{n-1}.

The following Python snippet realizes the steps in the computational algorithm.

```
# Given mesh points as arrays x and t (x[i], t[n])
dx = x[1] - x[0]
dt = t[1] - t[0]
C = c*dt/dx                # Courant number
Nt = len(t)-1
C2 = C**2                  # Help variable in the scheme

# Set initial condition u(x,0) = I(x)
for i in range(0, Nx+1):
    u_n[i] = I(x[i])

# Apply special formula for first step, incorporating du/dt=0
for i in range(1, Nx):
    u[i] = u_n[i] - \
           0.5*C**2(u_n[i+1] - 2*u_n[i] + u_n[i-1])
u[0] = 0;  u[Nx] = 0    # Enforce boundary conditions

# Switch variables before next step
u_nm1[:], u_n[:] = u_n, u

for n in range(1, Nt):
    # Update all inner mesh points at time t[n+1]
    for i in range(1, Nx):
        u[i] = 2u_n[i] - u_nm1[i] - \
               C**2(u_n[i+1] - 2*u_n[i] + u_n[i-1])

    # Insert boundary conditions
    u[0] = 0;  u[Nx] = 0

    # Switch variables before next step
    u_nm1[:], u_n[:] = u_n, u
```

[1] http://tinyurl.com/hbcasmj/book/html/mov-wave/D_stencil_gpl/index.html

[2] http://tinyurl.com/gokgkov/mov-wave/D_stencil_gpl/movie.ogg

2.2 Verification

Before implementing the algorithm, it is convenient to add a source term to the PDE (2.1), since that gives us more freedom in finding test problems for verification. Physically, a source term acts as a generator for waves in the interior of the domain.

2.2.1 A Slightly Generalized Model Problem

We now address the following extended initial-boundary value problem for one-dimensional wave phenomena:

$$u_{tt} = c^2 u_{xx} + f(x,t), \qquad x \in (0,L),\ t \in (0,T] \tag{2.17}$$

$$u(x,0) = I(x), \qquad x \in [0,L] \tag{2.18}$$

$$u_t(x,0) = V(x), \qquad x \in [0,L] \tag{2.19}$$

$$u(0,t) = 0, \qquad t > 0 \tag{2.20}$$

$$u(L,t) = 0, \qquad t > 0\,. \tag{2.21}$$

Sampling the PDE at (x_i, t_n) and using the same finite difference approximations as above, yields

$$[D_t D_t u = c^2 D_x D_x u + f]^n_i\,. \tag{2.22}$$

Writing this out and solving for the unknown u_i^{n+1} results in

$$u_i^{n+1} = -u_i^{n-1} + 2u_i^n + C^2(u_{i+1}^n - 2u_i^n + u_{i-1}^n) + \Delta t^2 f_i^n\,. \tag{2.23}$$

The equation for the first time step must be rederived. The discretization of the initial condition $u_t = V(x)$ at $t = 0$ becomes

$$[D_{2t} u = V]^0_i \quad \Rightarrow \quad u_i^{-1} = u_i^1 - 2\Delta t V_i,$$

which, when inserted in (2.23) for $n = 0$, gives the special formula

$$u_i^1 = u_i^0 - \Delta t V_i + \frac{1}{2}C^2\left(u_{i+1}^0 - 2u_i^0 + u_{i-1}^0\right) + \frac{1}{2}\Delta t^2 f_i^0\,. \tag{2.24}$$

2.2.2 Using an Analytical Solution of Physical Significance

Many wave problems feature sinusoidal oscillations in time and space. For example, the original PDE problem (2.1)–(2.5) allows an exact solution

$$u_e(x,t) = A\sin\left(\frac{\pi}{L}x\right)\cos\left(\frac{\pi}{L}ct\right)\,. \tag{2.25}$$

This u_e fulfills the PDE with $f = 0$, boundary conditions $u_e(0,t) = u_e(L,t) = 0$, as well as initial conditions $I(x) = A\sin\left(\frac{\pi}{L}x\right)$ and $V = 0$.

How to use exact solutions for verification
It is common to use such exact solutions of physical interest to verify implementations. However, the numerical solution u_i^n will only be an approximation to $u_e(x_i, t_n)$. We have no knowledge of the precise size of the error in this approximation, and therefore we can never know if discrepancies between u_i^n and $u_e(x_i, t_n)$ are caused by mathematical approximations or programming errors. In particular, if plots of the computed solution u_i^n and the exact one (2.25) look similar, many are tempted to claim that the implementation works. However, even if color plots look nice and the accuracy is "deemed good", there can still be serious programming errors present!

The only way to use exact physical solutions like (2.25) for serious and thorough verification is to run a series of simulations on finer and finer meshes, measure the integrated error in each mesh, and from this information estimate the empirical convergence rate of the method.

An introduction to the computing of convergence rates is given in Section 3.1.6 in [9]. There is also a detailed example on computing convergence rates in Sect. 1.2.2.

In the present problem, one expects the method to have a convergence rate of 2 (see Sect. 2.10), so if the computed rates are close to 2 on a sufficiently fine mesh, we have good evidence that the implementation is free of programming mistakes.

2.2.3 Manufactured Solution and Estimation of Convergence Rates

Specifying the solution and computing corresponding data One problem with the exact solution (2.25) is that it requires a simplification ($V = 0, f = 0$) of the implemented problem (2.17)–(2.21). An advantage of using a *manufactured solution* is that we can test all terms in the PDE problem. The idea of this approach is to set up some chosen solution and fit the source term, boundary conditions, and initial conditions to be compatible with the chosen solution. Given that our boundary conditions in the implementation are $u(0,t) = u(L,t) = 0$, we must choose a solution that fulfills these conditions. One example is

$$u_e(x,t) = x(L-x)\sin t\,.$$

Inserted in the PDE $u_{tt} = c^2 u_{xx} + f$ we get

$$-x(L-x)\sin t = -c^2 2\sin t + f \quad \Rightarrow f = (2c^2 - x(L-x))\sin t\,.$$

The initial conditions become

$$\begin{aligned} u(x,0) =& I(x) = 0,\\ u_t(x,0) =& V(x) = x(L-x)\,. \end{aligned}$$

Defining a single discretization parameter To verify the code, we compute the convergence rates in a series of simulations, letting each simulation use a finer mesh than the previous one. Such empirical estimation of convergence rates relies on an assumption that some measure E of the numerical error is related to the discretization parameters through

$$E = C_t\Delta t^r + C_x\Delta x^p,$$

where C_t, C_x, r, and p are constants. The constants r and p are known as the *convergence rates* in time and space, respectively. From the accuracy in the finite difference approximations, we expect $r = p = 2$, since the error terms are of order Δt^2 and Δx^2. This is confirmed by truncation error analysis and other types of analysis.

By using an exact solution of the PDE problem, we will next compute the error measure E on a sequence of refined meshes and see if the rates $r = p = 2$ are obtained. We will not be concerned with estimating the constants C_t and C_x, simply because we are not interested in their values.

It is advantageous to introduce a single discretization parameter $h = \Delta t = \hat{c}\Delta x$ for some constant $\hat{c}$. Since Δt and Δx are related through the Courant number, $\Delta t = C\Delta x/c$, we set $h = \Delta t$, and then $\Delta x = hc/C$. Now the expression for the error measure is greatly simplified:

$$E = C_t\Delta t^r + C_x\Delta x^r = C_t h^r + C_x\left(\frac{c}{C}\right)^r h^r = Dh^r,\quad D = C_t + C_x\left(\frac{c}{C}\right)^r.$$

Computing errors We choose an initial discretization parameter h_0 and run experiments with decreasing h: $h_i = 2^{-i}h_0$, $i = 1, 2, \ldots, m$. Halving h in each experiment is not necessary, but it is a common choice. For each experiment we must record E and h. Standard choices of error measure are the ℓ^2 and ℓ^∞ norms of the error mesh function e_i^n:

$$E = ||e_i^n||_{\ell^2} = \left(\Delta t\Delta x\sum_{n=0}^{N_t}\sum_{i=0}^{N_x}(e_i^n)^2\right)^{\frac{1}{2}},\quad e_i^n = u_e(x_i,t_n) - u_i^n, \tag{2.26}$$

$$E = ||e_i^n||_{\ell^\infty} = \max_{i,n}|e_i^n|. \tag{2.27}$$

In Python, one can compute $\sum_i (e_i^n)^2$ at each time step and accumulate the value in some sum variable, say `e2_sum`. At the final time step one can do `sqrt(dt*dx*e2_sum)`. For the ℓ^∞ norm one must compare the maximum error at a time level (`e.max()`) with the global maximum over the time domain: `e_max = max(e_max, e.max())`.

An alternative error measure is to use a spatial norm at one time step only, e.g., the end time T ($n = N_t$):

$$E = ||e_i^n||_{\ell^2} = \left(\Delta x\sum_{i=0}^{N_x}(e_i^n)^2\right)^{\frac{1}{2}},\quad e_i^n = u_e(x_i,t_n) - u_i^n, \tag{2.28}$$

$$E = ||e_i^n||_{\ell^\infty} = \max_{0\le i\le N_x}|e_i^n|. \tag{2.29}$$

The important point is that the error measure (E) for the simulation is represented by a single number.

Computing rates Let E_i be the error measure in experiment (mesh) number i (not to be confused with the spatial index i) and let h_i be the corresponding discretization parameter (h). With the error model $E_i = Dh_i^r$, we can estimate r by comparing two consecutive experiments:

$$E_{i+1} = Dh_{i+1}^r,$$
$$E_i = Dh_i^r.$$

Dividing the two equations eliminates the (uninteresting) constant D. Thereafter, solving for r yields

$$r = \frac{\ln E_{i+1}/E_i}{\ln h_{i+1}/h_i}.$$

Since r depends on i, i.e., which simulations we compare, we add an index to r: r_i, where $i = 0, \ldots, m-2$, if we have m experiments: $(h_0, E_0), \ldots, (h_{m-1}, E_{m-1})$.

In our present discretization of the wave equation we expect $r = 2$, and hence the r_i values should converge to 2 as i increases.

2.2.4 Constructing an Exact Solution of the Discrete Equations

With a manufactured or known analytical solution, as outlined above, we can estimate convergence rates and see if they have the correct asymptotic behavior. Experience shows that this is a quite good verification technique in that many common bugs will destroy the convergence rates. A significantly better test though, would be to check that the numerical solution is exactly what it should be. This will in general require exact knowledge of the numerical error, which we do not normally have (although we in Sect. 2.10 establish such knowledge in simple cases). However, it is possible to look for solutions where we can show that the numerical error vanishes, i.e., the solution of the original continuous PDE problem is also a solution of the discrete equations. This property often arises if the exact solution of the PDE is a lower-order polynomial. (Truncation error analysis leads to error measures that involve derivatives of the exact solution. In the present problem, the truncation error involves 4th-order derivatives of u in space and time. Choosing u as a polynomial of degree three or less will therefore lead to vanishing error.)

We shall now illustrate the construction of an exact solution to both the PDE itself and the discrete equations. Our chosen manufactured solution is quadratic in space and linear in time. More specifically, we set

$$u_e(x,t) = x(L-x)\left(1 + \frac{1}{2}t\right), \tag{2.30}$$

which by insertion in the PDE leads to $f(x,t) = 2(1+t)c^2$. This u_e fulfills the boundary conditions $u = 0$ and demands $I(x) = x(L-x)$ and $V(x) = \frac{1}{2}x(L-x)$.

To realize that the chosen u_e is also an exact solution of the discrete equations, we first remind ourselves that $t_n = n\Delta t$ so that

$$[D_tD_tt^2]^n = \frac{t_{n+1}^2 - 2t_n^2 + t_{n-1}^2}{\Delta t^2} = (n+1)^2 - 2n^2 + (n-1)^2 = 2, \tag{2.31}$$

$$[D_tD_tt]^n = \frac{t_{n+1} - 2t_n + t_{n-1}}{\Delta t^2} = \frac{((n+1) - 2n + (n-1))\Delta t}{\Delta t^2} = 0\,. \tag{2.32}$$

Hence,

$$[D_tD_tu_e]_i^n = x_i(L-x_i)\left[D_tD_t\left(1+\frac{1}{2}t\right)\right]^n = x_i(L-x_i)\frac{1}{2}[D_tD_tt]^n = 0\,.$$

Similarly, we get that

$$\begin{aligned}[D_xD_xu_e]_i^n &= \left(1+\frac{1}{2}t_n\right)[D_xD_x(xL-x^2)]_i\\ &= \left(1+\frac{1}{2}t_n\right)[LD_xD_xx - D_xD_xx^2]_i\\ &= -2\left(1+\frac{1}{2}t_n\right).\end{aligned}$$

Now, $f_i^n = 2(1+\frac{1}{2}t_n)c^2$, which results in

$$[D_tD_tu_e - c^2D_xD_xu_e - f]_i^n = 0 + c^2 2\left(1+\frac{1}{2}t_n\right) + 2\left(1+\frac{1}{2}t_n\right)c^2 = 0\,.$$

Moreover, $u_e(x_i, 0) = I(x_i)$, $\partial u_e/\partial t = V(x_i)$ at $t = 0$, and $u_e(x_0, t) = u_e(x_{N_x}, 0) = 0$. Also the modified scheme for the first time step is fulfilled by $u_e(x_i, t_n)$.

Therefore, the exact solution $u_e(x,t) = x(L-x)(1+t/2)$ of the PDE problem is also an exact solution of the discrete problem. This means that we know beforehand what numbers the numerical algorithm should produce. We can use this fact to check that the computed u_i^n values from an implementation equals $u_e(x_i, t_n)$, within machine precision. This result is valid *regardless of the mesh spacings* Δx and Δt! Nevertheless, there might be stability restrictions on Δx and Δt, so the test can only be run for a mesh that is compatible with the stability criterion (which in the present case is $C \leq 1$, to be derived later).

Notice

A product of quadratic or linear expressions in the various independent variables, as shown above, will often fulfill both the PDE problem and the discrete equations, and can therefore be very useful solutions for verifying implementations.

However, for 1D wave equations of the type $u_{tt} = c^2u_{xx}$ we shall see that there is always another much more powerful way of generating exact solutions (which consists in just setting $C = 1$ (!), as shown in Sect. 2.10).

2.3 Implementation

This section presents the complete computational algorithm, its implementation in Python code, animation of the solution, and verification of the implementation.

A real implementation of the basic computational algorithm from Sect. 2.1.5 and 2.1.6 can be encapsulated in a function, taking all the input data for the problem as arguments. The physical input data consists of c, $I(x)$, $V(x)$, $f(x,t)$, L, and T. The numerical input is the mesh parameters Δt and Δx.

Instead of specifying Δt *and* Δx, we can specify one of them and the Courant number C instead, since having explicit control of the Courant number is convenient when investigating the numerical method. Many find it natural to prescribe the resolution of the spatial grid and set N_x. The solver function can then compute $\Delta t = CL/(cN_x)$. However, for comparing $u(x,t)$ curves (as functions of x) for various Courant numbers it is more convenient to keep Δt fixed for all C and let Δx vary according to $\Delta x = c\Delta t/C$. With Δt fixed, all frames correspond to the same time t, and this simplifies animations that compare simulations with different mesh resolutions. Plotting functions of x with different spatial resolution is trivial, so it is easier to let Δx vary in the simulations than Δt.

2.3.1 Callback Function for User-Specific Actions

The solution at all spatial points at a new time level is stored in an array u of length $N_x + 1$. We need to decide what to do with this solution, e.g., visualize the curve, analyze the values, or write the array to file for later use. The decision about what to do is left to the user in the form of a user-supplied function

```
user_action(u, x, t, n)
```

where `u` is the solution at the spatial points `x` at time `t[n]`. The `user_action` function is called from the solver at each time level `n`.

If the user wants to plot the solution or store the solution at a time point, she needs to write such a function and take appropriate actions inside it. We will show examples on many such `user_action` functions.

Since the solver function makes calls back to the user's code via such a function, this type of function is called a *callback function*. When writing general software, like our solver function, which also needs to carry out special problem- or solution-dependent actions (like visualization), it is a common technique to leave those actions to user-supplied callback functions.

The callback function can be used to terminate the solution process if the user returns `True`. For example,

```
def my_user_action_function(u, x, t, n):
    return np.abs(u).max() > 10
```

is a callback function that will terminate the solver function (given below) of the amplitude of the waves exceed 10, which is here considered as a numerical instability.

2.3.2 The Solver Function

A first attempt at a solver function is listed below.

```
import numpy as np

def solver(I, V, f, c, L, dt, C, T, user_action=None):
    """Solve u_tt=c^2*u_xx + f on (0,L)x(0,T]."""
    Nt = int(round(T/dt))
    t = np.linspace(0, Nt*dt, Nt+1)    # Mesh points in time
    dx = dt*c/float(C)
    Nx = int(round(L/dx))
    x = np.linspace(0, L, Nx+1)        # Mesh points in space
    C2 = C**2                          # Help variable in the scheme
    # Make sure dx and dt are compatible with x and t
    dx = x[1] - x[0]
    dt = t[1] - t[0]

    if f is None or f == 0 :
        f = lambda x, t: 0
    if V is None or V == 0:
        V = lambda x: 0

    u     = np.zeros(Nx+1)   # Solution array at new time level
    u_n   = np.zeros(Nx+1)   # Solution at 1 time level back
    u_nm1 = np.zeros(Nx+1)   # Solution at 2 time levels back

    import time;  t0 = time.clock()  # Measure CPU time

    # Load initial condition into u_n
    for i in range(0,Nx+1):
        u_n[i] = I(x[i])

    if user_action is not None:
        user_action(u_n, x, t, 0)

    # Special formula for first time step
    n = 0
    for i in range(1, Nx):
        u[i] = u_n[i] + dt*V(x[i]) + \
               0.5*C2*(u_n[i-1] - 2*u_n[i] + u_n[i+1]) + \
               0.5*dt**2*f(x[i], t[n])
    u[0] = 0;  u[Nx] = 0

    if user_action is not None:
        user_action(u, x, t, 1)

    # Switch variables before next step
    u_nm1[:] = u_n;  u_n[:] = u
```

```
    for n in range(1, Nt):
        # Update all inner points at time t[n+1]
        for i in range(1, Nx):
            u[i] = - u_nm1[i] + 2*u_n[i] + \
                     C2*(u_n[i-1] - 2*u_n[i] + u_n[i+1]) + \
                     dt**2*f(x[i], t[n])

        # Insert boundary conditions
        u[0] = 0;  u[Nx] = 0
        if user_action is not None:
            if user_action(u, x, t, n+1):
                break


        # Switch variables before next step
        u_nm1[:] = u_n;  u_n[:] = u

    cpu_time = time.clock() - t0
    return u, x, t, cpu_time
```

A couple of remarks about the above code is perhaps necessary:

- Although we give `dt` and compute `dx` via `C` and `c`, the resulting `t` and `x` meshes do not necessarily correspond exactly to these values because of rounding errors. To explicitly ensure that `dx` and `dt` correspond to the cell sizes in `x` and `t`, we recompute the values.
- According to the particular choice made in Sect. 2.3.1, a true value returned from `user_action` should terminate the simulation. This is here implemented by a `break` statement inside the for loop in the solver.

2.3.3 Verification: Exact Quadratic Solution

We use the test problem derived in Sect. 2.2.1 for verification. Below is a unit test based on this test problem and realized as a proper *test function* compatible with the unit test frameworks nose or pytest.

```
def test_quadratic():
    """Check that u(x,t)=x(L-x)(1+t/2) is exactly reproduced."""

    def u_exact(x, t):
        return x*(L-x)*(1 + 0.5*t)

    def I(x):
        return u_exact(x, 0)

    def V(x):
        return 0.5*u_exact(x, 0)

    def f(x, t):
        return 2*(1 + 0.5*t)*c**2
```

```
    L = 2.5
    c = 1.5
    C = 0.75
    Nx = 6  # Very coarse mesh for this exact test
    dt = C*(L/Nx)/c
    T = 18

    def assert_no_error(u, x, t, n):
        u_e = u_exact(x, t[n])
        diff = np.abs(u - u_e).max()
        tol = 1E-13
        assert diff < tol

    solver(I, V, f, c, L, dt, C, T,
           user_action=assert_no_error)
```

When this function resides in the file `wave1D_u0.py`, one can run pytest to check that all test functions with names `test_*()` in this file work:

Terminal

```
Terminal> py.test -s -v wave1D_u0.py
```

2.3.4 Verification: Convergence Rates

A more general method, but not so reliable as a verification method, is to compute the convergence rates and see if they coincide with theoretical estimates. Here we expect a rate of 2 according to the various results in Sect. 2.10. A general function for computing convergence rates can be written like this:

```
def convergence_rates(
    u_exact,                  # Python function for exact solution
    I, V, f, c, L,            # physical parameters
    dt0, num_meshes, C, T):   # numerical parameters
    """
    Half the time step and estimate convergence rates for
    for num_meshes simulations.
    """
    # First define an appropriate user action function
    global error
    error = 0  # error computed in the user action function

    def compute_error(u, x, t, n):
        global error  # must be global to be altered here
        # (otherwise error is a local variable, different
        # from error defined in the parent function)
        if n == 0:
            error = 0
        else:
            error = max(error, np.abs(u - u_exact(x, t[n])).max())
```

```
    # Run finer and finer resolutions and compute true errors
    E = []
    h = []  # dt, solver adjusts dx such that C=dt*c/dx
    dt = dt0
    for i in range(num_meshes):
        solver(I, V, f, c, L, dt, C, T,
               user_action=compute_error)
        # error is computed in the final call to compute_error
        E.append(error)
        h.append(dt)
        dt /= 2  # halve the time step for next simulation
    print 'E:', E
    print 'h:', h
    # Convergence rates for two consecutive experiments
    r = [np.log(E[i]/E[i-1])/np.log(h[i]/h[i-1])
         for i in range(1,num_meshes)]
    return r
```

Using the analytical solution from Sect. 2.2.2, we can call `convergence_rates` to see if we get a convergence rate that approaches 2 and use the final estimate of the rate in an `assert` statement such that this function becomes a proper test function:

```
def test_convrate_sincos():
    n = m = 2
    L = 1.0
    u_exact = lambda x, t: np.cos(m*np.pi/L*t)*np.sin(m*np.pi/L*x)

    r = convergence_rates(
        u_exact=u_exact,
        I=lambda x: u_exact(x, 0),
        V=lambda x: 0,
        f=0,
        c=1,
        L=L,
        dt0=0.1,
        num_meshes=6,
        C=0.9,
        T=1)
    print 'rates sin(x)*cos(t) solution:', \
          [round(r_,2) for r_ in r]
    assert abs(r[-1] - 2) < 0.002
```

Doing `py.test -s -v wave1D_u0.py` will run also this test function and show the rates 2.05, 1.98, 2.00, 2.00, and 2.00 (to two decimals).

2.3.5 Visualization: Animating the Solution

Now that we have verified the implementation it is time to do a real computation where we also display evolution of the waves on the screen. Since the `solver` function knows nothing about what type of visualizations we may want, it calls the callback function `user_action(u, x, t, n)`. We must therefore write this function and find the proper statements for plotting the solution.

Function for administering the simulation The following `viz` function

1. defines a `user_action` callback function for plotting the solution at each time level,
2. calls the `solver` function, and
3. combines all the plots (in files) to video in different formats.

```
def viz(
    I, V, f, c, L, dt, C, T,  # PDE parameters
    umin, umax,               # Interval for u in plots
    animate=True,             # Simulation with animation?
    tool='matplotlib',        # 'matplotlib' or 'scitools'
    solver_function=solver,   # Function with numerical algorithm
    ):
    """Run solver and visualize u at each time level."""

    def plot_u_st(u, x, t, n):
        """user_action function for solver."""
        plt.plot(x, u, 'r-',
                 xlabel='x', ylabel='u',
                 axis=[0, L, umin, umax],
                 title='t=%f' % t[n], show=True)
        # Let the initial condition stay on the screen for 2
        # seconds, else insert a pause of 0.2 s between each plot
        time.sleep(2) if t[n] == 0 else time.sleep(0.2)
        plt.savefig('frame_%04d.png' % n)  # for movie making

    class PlotMatplotlib:
        def __call__(self, u, x, t, n):
            """user_action function for solver."""
            if n == 0:
                plt.ion()
                self.lines = plt.plot(x, u, 'r-')
                plt.xlabel('x');  plt.ylabel('u')
                plt.axis([0, L, umin, umax])
                plt.legend(['t=%f' % t[n]], loc='lower left')
            else:
                self.lines[0].set_ydata(u)
                plt.legend(['t=%f' % t[n]], loc='lower left')
                plt.draw()
            time.sleep(2) if t[n] == 0 else time.sleep(0.2)
            plt.savefig('tmp_%04d.png' % n)  # for movie making

    if tool == 'matplotlib':
        import matplotlib.pyplot as plt
        plot_u = PlotMatplotlib()
    elif tool == 'scitools':
        import scitools.std as plt  # scitools.easyviz interface
        plot_u = plot_u_st
    import time, glob, os

    # Clean up old movie frames
    for filename in glob.glob('tmp_*.png'):
        os.remove(filename)
```

```
    # Call solver and do the simulation
    user_action = plot_u if animate else None
    u, x, t, cpu = solver_function(
        I, V, f, c, L, dt, C, T, user_action)

    # Make video files
    fps = 4  # frames per second
    codec2ext = dict(flv='flv', libx264='mp4', libvpx='webm',
                     libtheora='ogg')  # video formats
    filespec = 'tmp_%04d.png'
    movie_program = 'ffmpeg'  # or 'avconv'
    for codec in codec2ext:
        ext = codec2ext[codec]
        cmd = '%(movie_program)s -r %(fps)d -i %(filespec)s '\
              '-vcodec %(codec)s movie.%(ext)s' % vars()
        os.system(cmd)

    if tool == 'scitools':
        # Make an HTML play for showing the animation in a browser
        plt.movie('tmp_*.png', encoder='html', fps=fps,
                  output_file='movie.html')
    return cpu
```

Dissection of the code The viz function can either use SciTools or Matplotlib for visualizing the solution. The user_action function based on SciTools is called plot_u_st, while the user_action function based on Matplotlib is a bit more complicated as it is realized as a class and needs statements that differ from those for making static plots. SciTools can utilize both Matplotlib and Gnuplot (and many other plotting programs) for doing the graphics, but Gnuplot is a relevant choice for large N_x or in two-dimensional problems as Gnuplot is significantly faster than Matplotlib for screen animations.

A function inside another function, like plot_u_st in the above code segment, has access to *and remembers* all the local variables in the surrounding code inside the viz function (!). This is known in computer science as a *closure* and is very convenient to program with. For example, the plt and time modules defined outside plot_u are accessible for plot_u_st when the function is called (as user_action) in the solver function. Some may think, however, that a class instead of a closure is a cleaner and easier-to-understand implementation of the user action function, see Sect. 2.8.

The plot_u_st function just makes a standard SciTools plot command for plotting u as a function of x at time t[n]. To achieve a smooth animation, the plot command should take keyword arguments instead of being broken into separate calls to xlabel, ylabel, axis, time, and show. Several plot calls will automatically cause an animation on the screen. In addition, we want to save each frame in the animation to file. We then need a filename where the frame number is padded with zeros, here tmp_0000.png, tmp_0001.png, and so on. The proper printf construction is then tmp_%04d.png. Section 1.3.2 contains more basic information on making animations.

The solver is called with an argument plot_u as user_function. If the user chooses to use SciTools, plot_u is the plot_u_st callback function, but for Matplotlib it is an instance of the class PlotMatplotlib. Also this class makes use of variables defined in the viz function: plt and time. With Matplotlib, one has to make the first plot the standard way, and then update the y data in the plot at every time level. The update requires active use of the returned value from plt.plot in the first plot. This value would need to be stored in a local variable if we were to use a closure for the user_action function when doing the animation with Matplotlib. It is much easier to store the variable as a class attribute self.lines. Since the class is essentially a function, we implement the function as the special method __call__ such that the instance plot_u(u, x, t, n) can be called as a standard callback function from solver.

Making movie files From the frame_*.png files containing the frames in the animation we can make video files. Section 1.3.2 presents basic information on how to use the ffmpeg (or avconv) program for producing video files in different modern formats: Flash, MP4, Webm, and Ogg.

The viz function creates an ffmpeg or avconv command with the proper arguments for each of the formats Flash, MP4, WebM, and Ogg. The task is greatly simplified by having a codec2ext dictionary for mapping video codec names to filename extensions. As mentioned in Sect. 1.3.2, only two formats are actually needed to ensure that all browsers can successfully play the video: MP4 and WebM.

Some animations having a large number of plot files may not be properly combined into a video using ffmpeg or avconv. A method that always works is to play the PNG files as an animation in a browser using JavaScript code in an HTML file. The SciTools package has a function movie (or a stand-alone command scitools movie) for creating such an HTML player. The plt.movie call in the viz function shows how the function is used. The file movie.html can be loaded into a browser and features a user interface where the speed of the animation can be controlled. Note that the movie in this case consists of the movie.html file and all the frame files tmp_*.png.

Skipping frames for animation speed Sometimes the time step is small and T is large, leading to an inconveniently large number of plot files and a slow animation on the screen. The solution to such a problem is to decide on a total number of frames in the animation, num_frames, and plot the solution only for every skip_frame frames. For example, setting skip_frame=5 leads to plots of every 5 frames. The default value skip_frame=1 plots every frame. The total number of time levels (i.e., maximum possible number of frames) is the length of t, t.size (or len(t)), so if we want num_frames frames in the animation, we need to plot every t.size/num_frames frames:

```
skip_frame = int(t.size/float(num_frames))
if n % skip_frame == 0 or n == t.size-1:
    st.plot(x, u, 'r-', ...)
```

The initial condition (n=0) is included by n % skip_frame == 0, as well as every skip_frame-th frame. As n % skip_frame == 0 will very seldom be true for the very final frame, we must also check if n == t.size-1 to get the final frame included.

A simple choice of numbers may illustrate the formulas: say we have 801 frames in total (`t.size`) and we allow only 60 frames to be plotted. As `n` then runs from 801 to 0, we need to plot every 801/60 frame, which with integer division yields 13 as `skip_frame`. Using the mod function, `n % skip_frame`, this operation is zero every time `n` can be divided by 13 without a remainder. That is, the `if` test is true when `n` equals $0, 13, 26, 39, \ldots, 780, 801$. The associated code is included in the `plot_u` function, inside the `viz` function, in the file `wave1D_u0.py`.

2.3.6 Running a Case

The first demo of our 1D wave equation solver concerns vibrations of a string that is initially deformed to a triangular shape, like when picking a guitar string:

$$I(x) = \begin{cases} ax/x_0, & x < x_0, \\ a(L-x)/(L-x_0), & \text{otherwise} \end{cases} . \tag{2.33}$$

We choose $L = 75\,\text{cm}$, $x_0 = 0.8L$, $a = 5$ mm, and a time frequency $\nu = 440\,\text{Hz}$. The relation between the wave speed c and ν is $c = \nu\lambda$, where λ is the wavelength, taken as $2L$ because the longest wave on the string forms half a wavelength. There is no external force, so $f = 0$ (meaning we can neglect gravity), and the string is at rest initially, implying $V = 0$.

Regarding numerical parameters, we need to specify a Δt. Sometimes it is more natural to think of a spatial resolution instead of a time step. A natural semi-coarse spatial resolution in the present problem is $N_x = 50$. We can then choose the associated Δt (as required by the `viz` and `solver` functions) as the stability limit: $\Delta t = L/(N_x c)$. This is the Δt to be specified, but notice that if $C < 1$, the actual Δx computed in `solver` gets larger than L/N_x: $\Delta x = c\Delta t/C = L/(N_x C)$. (The reason is that we fix Δt and adjust Δx, so if C gets smaller, the code implements this effect in terms of a larger Δx.)

A function for setting the physical and numerical parameters and calling `viz` in this application goes as follows:

```
def guitar(C):
    """Triangular wave (pulled guitar string)."""
    L = 0.75
    x0 = 0.8*L
    a = 0.005
    freq = 440
    wavelength = 2*L
    c = freq*wavelength
    omega = 2*pi*freq
    num_periods = 1
    T = 2*pi/omega*num_periods
    # Choose dt the same as the stability limit for Nx=50
    dt = L/50./c

    def I(x):
        return a*x/x0 if x < x0 else a/(L-x0)*(L-x)

    umin = -1.2*a;  umax = -umin
    cpu = viz(I, 0, 0, c, L, dt, C, T, umin, umax,
              animate=True, tool='scitools')
```

The associated program has the name `wave1D_u0.py`. Run the program and watch the movie of the vibrating string[3]. The string should ideally consist of straight segments, but these are somewhat wavy due to numerical approximation. Run the case with the `wave1D_u0.py` code and $C = 1$ to see the exact solution.

2.3.7 Working with a Scaled PDE Model

Depending on the model, it may be a substantial job to establish consistent and relevant physical parameter values for a case. The guitar string example illustrates the point. However, by *scaling* the mathematical problem we can often reduce the need to estimate physical parameters dramatically. The scaling technique consists of introducing new independent and dependent variables, with the aim that the absolute values of these lie in $[0, 1]$. We introduce the dimensionless variables (details are found in Section 3.1.1 in [11])

$$\bar{x} = \frac{x}{L}, \quad \bar{t} = \frac{c}{L}t, \quad \bar{u} = \frac{u}{a}.$$

Here, L is a typical length scale, e.g., the length of the domain, and a is a typical size of u, e.g., determined from the initial condition: $a = \max_x |I(x)|$.

We get by the chain rule that

$$\frac{\partial u}{\partial t} = \frac{\partial}{\partial \bar{t}}(a\bar{u})\frac{d\bar{t}}{dt} = \frac{ac}{L}\frac{\partial \bar{u}}{\partial \bar{t}}.$$

Similarly,

$$\frac{\partial u}{\partial x} = \frac{a}{L}\frac{\partial \bar{u}}{\partial \bar{x}}.$$

Inserting the dimensionless variables in the PDE gives, in case $f = 0$,

$$\frac{a^2c^2}{L^2}\frac{\partial^2 \bar{u}}{\partial \bar{t}^2} = \frac{a^2c^2}{L^2}\frac{\partial^2 \bar{u}}{\partial \bar{x}^2}.$$

Dropping the bars, we arrive at the scaled PDE

$$\frac{\partial^2 u}{\partial t^2} = \frac{\partial^2 u}{\partial x^2}, \quad x \in (0, 1),\ t \in (0, cT/L), \tag{2.34}$$

which has no parameter c^2 anymore. The initial conditions are scaled as

$$a\bar{u}(\bar{x}, 0) = I(L\bar{x})$$

and

$$\frac{a}{L/c}\frac{\partial \bar{u}}{\partial \bar{t}}(\bar{x}, 0) = V(L\bar{x}),$$

[3] http://tinyurl.com/hbcasmj/wave/html/mov-wave/guitar_C0.8/movie.html

resulting in

$$\bar{u}(\bar{x},0) = \frac{I(L\bar{x})}{\max_x |I(x)|}, \quad \frac{\partial \bar{u}}{\partial \bar{t}}(\bar{x},0) = \frac{L}{ac}V(L\bar{x}).$$

In the common case $V = 0$ we see that there are no physical parameters to be estimated in the PDE model!

If we have a program implemented for the physical wave equation with dimensions, we can obtain the dimensionless, scaled version by setting $c = 1$. The initial condition of a guitar string, given in (2.33), gets its scaled form by choosing $a = 1$, $L = 1$, and $x_0 \in [0, 1]$. This means that we only need to decide on the x_0 value as a fraction of unity, because the scaled problem corresponds to setting all other parameters to unity. In the code we can just set `a=c=L=1, x0=0.8`, and there is no need to calculate with wavelengths and frequencies to estimate c!

The only non-trivial parameter to estimate in the scaled problem is the final end time of the simulation, or more precisely, how it relates to periods in periodic solutions in time, since we often want to express the end time as a certain number of periods. The period in the dimensionless problem is 2, so the end time can be set to the desired number of periods times 2.

Why the dimensionless period is 2 can be explained by the following reasoning. Suppose that u behaves as $\cos(\omega t)$ in time in the original problem with dimensions. The corresponding period is then $P = 2\pi/\omega$, but we need to estimate ω. A typical solution of the wave equation is $u(x,t) = A\cos(kx)\cos(\omega t)$, where A is an amplitude and k is related to the wave length λ in space: $\lambda = 2\pi/k$. Both λ and A will be given by the initial condition $I(x)$. Inserting this $u(x,t)$ in the PDE yields $-\omega^2 = -c^2k^2$, i.e., $\omega = kc$. The period is therefore $P = 2\pi/(kc)$. If the boundary conditions are $u(0,t) = u(L,t)$, we need to have $kL = n\pi$ for integer n. The period becomes $P = 2L/nc$. The longest period is $P = 2L/c$. The dimensionless period $\tilde{P}$ is obtained by dividing P by the time scale L/c, which results in $\tilde{P} = 2$. Shorter waves in the initial condition will have a dimensionless shorter period $\tilde{P} = 2/n$ $(n > 1)$.

2.4 Vectorization

The computational algorithm for solving the wave equation visits one mesh point at a time and evaluates a formula for the new value u_i^{n+1} at that point. Technically, this is implemented by a loop over array elements in a program. Such loops may run slowly in Python (and similar interpreted languages such as R and MATLAB). One technique for speeding up loops is to perform operations on entire arrays instead of working with one element at a time. This is referred to as *vectorization*, *vector computing*, or *array computing*. Operations on whole arrays are possible if the computations involving each element is independent of each other and therefore can, at least in principle, be performed simultaneously. That is, vectorization not only speeds up the code on serial computers, but also makes it easy to exploit parallel computing. Actually, there are Python tools like Numba[4] that can automatically turn vectorized code into parallel code.

[4] http://numba.pydata.org

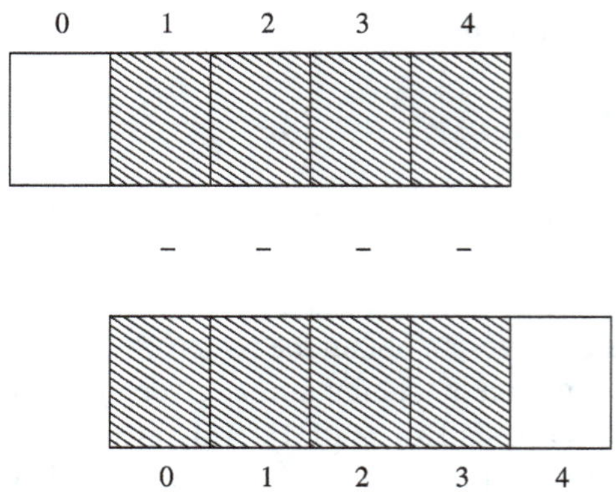

Fig. 2.3 Illustration of subtracting two slices of two arrays

2.4.1 Operations on Slices of Arrays

Efficient computing with numpy arrays demands that we avoid loops and compute with entire arrays at once (or at least large portions of them). Consider this calculation of differences $d_i = u_{i+1} - u_i$:

```
n = u.size
for i in range(0, n-1):
    d[i] = u[i+1] - u[i]
```

All the differences here are independent of each other. The computation of d can therefore alternatively be done by subtracting the array $(u_0, u_1, \ldots, u_{n-1})$ from the array where the elements are shifted one index upwards: $(u_1, u_2, \ldots, u_n)$, see Fig. 2.3. The former subset of the array can be expressed by u[0:n-1], u[0:-1], or just u[:-1], meaning from index 0 up to, but not including, the last element (-1). The latter subset is obtained by u[1:n] or u[1:], meaning from index 1 and the rest of the array. The computation of d can now be done without an explicit Python loop:

```
d = u[1:] - u[:-1]
```

or with explicit limits if desired:

```
d = u[1:n] - u[0:n-1]
```

Indices with a colon, going from an index to (but not including) another index are called *slices*. With numpy arrays, the computations are still done by loops, but in efficient, compiled, highly optimized C or Fortran code. Such loops are sometimes referred to as *vectorized loops*. Such loops can also easily be distributed among many processors on parallel computers. We say that the *scalar code* above, working on an element (a scalar) at a time, has been replaced by an equivalent *vectorized code*. The process of vectorizing code is called *vectorization*.

Test your understanding

Newcomers to vectorization are encouraged to choose a small array u, say with five elements, and simulate with pen and paper both the loop version and the vectorized version above.

Finite difference schemes basically contain differences between array elements with shifted indices. As an example, consider the updating formula

```
for i in range(1, n-1):
    u2[i] = u[i-1] - 2*u[i] + u[i+1]
```

The vectorization consists of replacing the loop by arithmetics on slices of arrays of length n-2:

```
u2 = u[:-2] - 2*u[1:-1] + u[2:]
u2 = u[0:n-2] - 2*u[1:n-1] + u[2:n]    # alternative
```

Note that the length of u2 becomes n-2. If u2 is already an array of length n and we want to use the formula to update all the "inner" elements of u2, as we will when solving a 1D wave equation, we can write

```
u2[1:-1]  = u[:-2] - 2*u[1:-1] + u[2:]
u2[1:n-1] = u[0:n-2] - 2*u[1:n-1] + u[2:n]    # alternative
```

The first expression's right-hand side is realized by the following steps, involving temporary arrays with intermediate results, since each array operation can only involve one or two arrays. The numpy package performs (behind the scenes) the first line above in four steps:

```
temp1 = 2*u[1:-1]
temp2 = u[:-2] - temp1
temp3 = temp2 + u[2:]
u2[1:-1] = temp3
```

We need three temporary arrays, but a user does not need to worry about such temporary arrays.

Common mistakes with array slices

Array expressions with slices demand that the slices have the same shape. It easy to make a mistake in, e.g.,

```
u2[1:n-1] = u[0:n-2] - 2*u[1:n-1] + u[2:n]
```

and write

```
u2[1:n-1] = u[0:n-2] - 2*u[1:n-1] + u[1:n]
```

Now `u[1:n]` has wrong length (`n-1`) compared to the other array slices, causing a `ValueError` and the message `could not broadcast input array from shape 103 into shape 104` (if `n` is 105). When such errors occur one must closely examine all the slices. Usually, it is easier to get upper limits of slices right when they use `-1` or `-2` or empty limit rather than expressions involving the length.

Another common mistake, when u2 has length n, is to forget the slice in the array on the left-hand side,

```
u2 = u[0:n-2] - 2*u[1:n-1] + u[1:n]
```

This is really crucial: now u2 becomes a *new* array of length `n-2`, which is the wrong length as we have no entries for the boundary values. We meant to insert the right-hand side array *into* the original u2 array for the entries that correspond to the internal points in the mesh (`1:n-1` or `1:-1`).

Vectorization may also work nicely with functions. To illustrate, we may extend the previous example as follows:

```
def f(x):
    return x**2 + 1

for i in range(1, n-1):
    u2[i] = u[i-1] - 2*u[i] + u[i+1] + f(x[i])
```

Assuming u2, u, and x all have length n, the vectorized version becomes

```
u2[1:-1] = u[:-2] - 2*u[1:-1] + u[2:] + f(x[1:-1])
```

Obviously, `f` must be able to take an array as argument for `f(x[1:-1])` to make sense.

2.4.2 Finite Difference Schemes Expressed as Slices

We now have the necessary tools to vectorize the wave equation algorithm as described mathematically in Sect. 2.1.5 and through code in Sect. 2.3.2. There are three loops: one for the initial condition, one for the first time step, and finally the loop that is repeated for all subsequent time levels. Since only the latter is repeated a potentially large number of times, we limit our vectorization efforts to this loop. Within the time loop, the space loop reads:

```
for i in range(1, Nx):
    u[i] = 2*u_n[i] - u_nm1[i] + \
           C2*(u_n[i-1] - 2*u_n[i] + u_n[i+1])
```

The vectorized version becomes

```
u[1:-1] = - u_nm1[1:-1] + 2*u_n[1:-1] + \
          C2*(u_n[:-2] - 2*u_n[1:-1] + u_n[2:])
```

or

```
u[1:Nx] = 2*u_n[1:Nx]- u_nm1[1:Nx] + \
          C2*(u_n[0:Nx-1] - 2*u_n[1:Nx] + u_n[2:Nx+1])
```

The program wave1D_u0v.py contains a new version of the function solver where both the scalar and the vectorized loops are included (the argument version is set to scalar or vectorized, respectively).

2.4.3 Verification

We may reuse the quadratic solution $u_e(x,t) = x(L-x)(1+\frac{1}{2}t)$ for verifying also the vectorized code. A test function can now verify both the scalar and the vectorized version. Moreover, we may use a user_action function that compares the computed and exact solution at each time level and performs a test:

```
def test_quadratic():
    """
    Check the scalar and vectorized versions for
    a quadratic u(x,t)=x(L-x)(1+t/2) that is exactly reproduced.
    """
    # The following function must work for x as array or scalar
    u_exact = lambda x, t: x*(L - x)*(1 + 0.5*t)
    I = lambda x: u_exact(x, 0)
    V = lambda x: 0.5*u_exact(x, 0)
    # f is a scalar (zeros_like(x) works for scalar x too)
    f = lambda x, t: np.zeros_like(x) + 2*c**2*(1 + 0.5*t)

    L = 2.5
    c = 1.5
    C = 0.75
    Nx = 3  # Very coarse mesh for this exact test
    dt = C*(L/Nx)/c
    T = 18

    def assert_no_error(u, x, t, n):
        u_e = u_exact(x, t[n])
        tol = 1E-13
        diff = np.abs(u - u_e).max()
        assert diff < tol

    solver(I, V, f, c, L, dt, C, T,
           user_action=assert_no_error, version='scalar')
    solver(I, V, f, c, L, dt, C, T,
           user_action=assert_no_error, version='vectorized')
```

Lambda functions

The code segment above demonstrates how to achieve very compact code, without degraded readability, by use of lambda functions for the various input parameters that require a Python function. In essence,

```
f = lambda x, t: L*(x-t)**2
```

is equivalent to

```
def f(x, t):
    return L(x-t)**2
```

Note that lambda functions can just contain a single expression and no statements.

One advantage with lambda functions is that they can be used directly in calls:

```
solver(I=lambda x: sin(pi*x/L), V=0, f=0, ...)
```

2.4.4 Efficiency Measurements

The `wave1D_u0v.py` contains our new `solver` function with both scalar and vectorized code. For comparing the efficiency of scalar versus vectorized code, we need a `viz` function as discussed in Sect. 2.3.5. All of this `viz` function can be reused, except the call to `solver_function`. This call lacks the parameter `version`, which we want to set to `vectorized` and `scalar` for our efficiency measurements.

One solution is to copy the `viz` code from `wave1D_u0` into `wave1D_u0v.py` and add a `version` argument to the `solver_function` call. Taking into account how much animation code we then duplicate, this is not a good idea. Alternatively, introducing the `version` argument in `wave1D_u0.viz`, so that this function can be imported into `wave1D_u0v.py`, is not a good solution either, since `version` has no meaning in that file. We need better ideas!

Solution 1 Calling `viz` in `wave1D_u0` with `solver_function` as our new solver in `wave1D_u0v` works fine, since this solver has `version='vectorized'` as default value. The problem arises when we want to test `version='scalar'`. The simplest solution is then to use `wave1D_u0.solver` instead. We make a new `viz` function in `wave1D_u0v.py` that has a `version` argument and that just calls `wave1D_u0.viz`:

```
def viz(
    I, V, f, c, L, dt, C, T,  # PDE parameters
    umin, umax,               # Interval for u in plots
    animate=True,             # Simulation with animation?
    tool='matplotlib',        # 'matplotlib' or 'scitools'
    solver_function=solver,   # Function with numerical algorithm
    version='vectorized',     # 'scalar' or 'vectorized'
    ):
    import wave1D_u0
    if version == 'vectorized':
        # Reuse viz from wave1D_u0, but with the present
        # modules' new vectorized solver (which has
        # version='vectorized' as default argument;
        # wave1D_u0.viz does not feature this argument)
        cpu = wave1D_u0.viz(
            I, V, f, c, L, dt, C, T, umin, umax,
            animate, tool, solver_function=solver)
    elif version == 'scalar':
        # Call wave1D_u0.viz with a solver with
        # scalar code and use wave1D_u0.solver.
        cpu = wave1D_u0.viz(
            I, V, f, c, L, dt, C, T, umin, umax,
            animate, tool,
            solver_function=wave1D_u0.solver)
```

Solution 2 There is a more advanced and fancier solution featuring a very useful trick: we can make a new function that will always call `wave1D_u0v.solver` with `version='scalar'`. The `functools.partial` function from standard Python takes a function `func` as argument and a series of positional and keyword arguments and returns a new function that will call `func` with the supplied arguments, while the user can control all the other arguments in `func`. Consider a trivial example,

```
def f(a, b, c=2):
    return a + b + c
```

We want to ensure that `f` is always called with `c=3`, i.e., `f` has only two "free" arguments `a` and `b`. This functionality is obtained by

```
import functools
f2 = functools.partial(f, c=3)

print f2(1, 2)  # results in 1+2+3=6
```

Now `f2` calls `f` with whatever the user supplies as `a` and `b`, but `c` is always 3.

Back to our `viz` code, we can do

```
import functools
# Call wave1D_u0.solver with version fixed to scalar
scalar_solver = functools.partial(wave1D_u0.solver, version='scalar')
cpu = wave1D_u0.viz(
        I, V, f, c, L, dt, C, T, umin, umax,
        animate, tool, solver_function=scalar_solver)
```

The new `scalar_solver` takes the same arguments as `wave1D_u0.scalar` and calls `wave1D_u0v.scalar`, but always supplies the extra argument `version='scalar'`. When sending this `solver_function` to `wave1D_u0.viz`, the latter will call `wave1D_u0v.solver` with all the `I`, `V`, `f`, etc., arguments we supply, plus `version='scalar'`.

Efficiency experiments We now have a `viz` function that can call our solver function both in scalar and vectorized mode. The function `run_efficiency_experiments` in `wave1D_u0v.py` performs a set of experiments and reports the CPU time spent in the scalar and vectorized solver for the previous string vibration example with spatial mesh resolutions $N_x = 50, 100, 200, 400, 800$. Running this function reveals that the vectorized code runs substantially faster: the vectorized code runs approximately $N_x/10$ times as fast as the scalar code!

2.4.5 Remark on the Updating of Arrays

At the end of each time step we need to update the `u_nm1` and `u_n` arrays such that they have the right content for the next time step:

```
u_nm1[:] = u_n
u_n[:] = u
```

The order here is important: updating `u_n` first, makes `u_nm1` equal to `u`, which is wrong!

The assignment `u_n[:] = u` copies the content of the `u` array into the elements of the `u_n` array. Such copying takes time, but that time is negligible compared to the time needed for computing `u` from the finite difference formula, even when the formula has a vectorized implementation. However, efficiency of program code is a key topic when solving PDEs numerically (particularly when there are two or three space dimensions), so it must be mentioned that there exists a much more efficient way of making the arrays `u_nm1` and `u_n` ready for the next time step. The idea is based on *switching references* and explained as follows.

A Python variable is actually a reference to some object (C programmers may think of pointers). Instead of copying data, we can let `u_nm1` refer to the `u_n` object and `u_n` refer to the `u` object. This is a very efficient operation (like switching pointers in C). A naive implementation like

```
u_nm1 = u_n
u_n = u
```

will fail, however, because now `u_nm1` refers to the `u_n` object, but then the name `u_n` refers to `u`, so that this `u` object has two references, `u_n` and `u`, while our third array, originally referred to by `u_nm1`, has no more references and is lost. This means that the variables `u`, `u_n`, and `u_nm1` refer to two arrays and not three. Consequently, the computations at the next time level will be messed up, since updating the elements in u will imply updating the elements in u_n too, thereby destroying the solution at the previous time step.

While `u_nm1 = u_n` is fine, `u_n = u` is problematic, so the solution to this problem is to ensure that u points to the u_nm1 array. This is mathematically wrong, but new correct values will be filled into u at the next time step and make it right.

The correct switch of references is

```
tmp = u_nm1
u_nm1 = u_n
u_n = u
u = tmp
```

We can get rid of the temporary reference `tmp` by writing

```
u_nm1, u_n, u = u_n, u, u_nm1
```

This switching of references for updating our arrays will be used in later implementations.

Caution

The update `u_nm1, u_n, u = u_n, u, u_nm1` leaves wrong content in u at the final time step. This means that if we return u, as we do in the example codes here, we actually return u_nm1, which is obviously wrong. It is therefore important to adjust the content of u to `u = u_n` before returning u. (Note that the `user_action` function reduces the need to return the solution from the solver.)

2.5 Exercises

Exercise 2.1: Simulate a standing wave

The purpose of this exercise is to simulate standing waves on $[0, L]$ and illustrate the error in the simulation. Standing waves arise from an initial condition

$$u(x,0) = A \sin\left(\frac{\pi}{L}mx\right),$$

where m is an integer and A is a freely chosen amplitude. The corresponding exact solution can be computed and reads

$$u_e(x,t) = A \sin\left(\frac{\pi}{L}mx\right) \cos\left(\frac{\pi}{L}mct\right).$$

a) Explain that for a function $\sin kx \cos \omega t$ the wave length in space is $\lambda = 2\pi/k$ and the period in time is $P = 2\pi/\omega$. Use these expressions to find the wave length in space and period in time of u_e above.
b) Import the `solver` function from `wave1D_u0.py` into a new file where the `viz` function is reimplemented such that it plots either the numerical *and* the exact solution, *or* the error.
c) Make animations where you illustrate how the error $e_i^n = u_e(x_i, t_n) - u_i^n$ develops and increases in time. Also make animations of u and u_e simultaneously.

Hint 1 Quite long time simulations are needed in order to display significant discrepancies between the numerical and exact solution.

Hint 2 A possible set of parameters is $L = 12$, $m = 9$, $c = 2$, $A = 1$, $N_x = 80$, $C = 0.8$. The error mesh function e^n can be simulated for 10 periods, while 20–30 periods are needed to show significant differences between the curves for the numerical and exact solution.
Filename: `wave_standing`.

Remarks The important parameters for numerical quality are C and $k\Delta x$, where $C = c\Delta t/\Delta x$ is the Courant number and k is defined above ($k\Delta x$ is proportional to how many mesh points we have per wave length in space, see Sect. 2.10.4 for explanation).

Exercise 2.2: Add storage of solution in a user action function
Extend the `plot_u` function in the file `wave1D_u0.py` to also store the solutions `u` in a list. To this end, declare `all_u` as an empty list in the `viz` function, outside `plot_u`, and perform an append operation inside the `plot_u` function. Note that a function, like `plot_u`, inside another function, like `viz`, remembers all local variables in `viz` function, including `all_u`, even when `plot_u` is called (as `user_action`) in the `solver` function. Test both `all_u.append(u)` and `all_u.append(u.copy())`. Why does one of these constructions fail to store the solution correctly? Let the `viz` function return the `all_u` list converted to a two-dimensional `numpy` array.
Filename: `wave1D_u0_s_store`.

Exercise 2.3: Use a class for the user action function
Redo Exercise 2.2 using a class for the user action function. Let the `all_u` list be an attribute in this class and implement the user action function as a method (the special method `__call__` is a natural choice). The class versions avoid that the user action function depends on parameters defined outside the function (such as `all_u` in Exercise 2.2).
Filename: `wave1D_u0_s2c`.

Exercise 2.4: Compare several Courant numbers in one movie
The goal of this exercise is to make movies where several curves, corresponding to different Courant numbers, are visualized. Write a program that resembles `wave1D_u0_s2c.py` in Exercise 2.3, but with a `viz` function that can take a list of `C` values as argument and create a movie with solutions corresponding to the given `C` values. The `plot_u` function must be changed to store the solution in an array (see Exercise 2.2 or 2.3 for details), `solver` must be computed for each value of the Courant number, and finally one must run through each time step and plot all the spatial solution curves in one figure and store it in a file.

The challenge in such a visualization is to ensure that the curves in one plot correspond to the same time point. The easiest remedy is to keep the time resolution constant and change the space resolution to change the Courant number. Note that each spatial grid is needed for the final plotting, so it is an option to store those grids too.
Filename: `wave_numerics_comparison`.

Exercise 2.5: Implementing the solver function as a generator
The callback function user_action(u, x, t, n) is called from the solver function (in, e.g., wave1D_u0.py) at every time level and lets the user work perform desired actions with the solution, like plotting it on the screen. We have implemented the callback function in the typical way it would have been done in C and Fortran. Specifically, the code looks like

```
if user_action is not None:
    if user_action(u, x, t, n):
        break
```

Many Python programmers, however, may claim that solver is an iterative process, and that iterative processes with callbacks to the user code is more elegantly implemented as *generators*. The rest of the text has little meaning unless you are familiar with Python generators and the yield statement.

Instead of calling user_action, the solver function issues a yield statement, which is a kind of return statement:

```
yield u, x, t, n
```

The program control is directed back to the calling code:

```
for u, x, t, n in solver(...):
    # Do something with u at t[n]
```

When the block is done, solver continues with the statement after yield. Note that the functionality of terminating the solution process if user_action returns a True value is not possible to implement in the generator case.

Implement the solver function as a generator, and plot the solution at each time step.
Filename: wave1D_u0_generator.

Project 2.6: Calculus with 1D mesh functions
This project explores integration and differentiation of mesh functions, both with scalar and vectorized implementations. We are given a mesh function f_i on a spatial one-dimensional mesh $x_i = i\Delta x, i = 0, \ldots, N_x$, over the interval $[a, b]$.

a) Define the discrete derivative of f_i by using centered differences at internal mesh points and one-sided differences at the end points. Implement a scalar version of the computation in a Python function and write an associated unit test for the linear case $f(x) = 4x - 2.5$ where the discrete derivative should be exact.
b) Vectorize the implementation of the discrete derivative. Extend the unit test to check the validity of the implementation.
c) To compute the discrete integral F_i of f_i, we assume that the mesh function f_i varies linearly between the mesh points. Let $f(x)$ be such a linear interpolant

of f_i. We then have

$$F_i = \int_{x_0}^{x_i} f(x)dx\,.$$

The exact integral of a piecewise linear function $f(x)$ is given by the Trapezoidal rule. Show that if F_i is already computed, we can find F_{i+1} from

$$F_{i+1} = F_i + \frac{1}{2}(f_i + f_{i+1})\Delta x\,.$$

Make a function for the scalar implementation of the discrete integral as a mesh function. That is, the function should return F_i for $i = 0, \ldots, N_x$. For a unit test one can use the fact that the above defined discrete integral of a linear function (say $f(x) = 4x - 2.5$) is exact.

d) Vectorize the implementation of the discrete integral. Extend the unit test to check the validity of the implementation.

Hint Interpret the recursive formula for F_{i+1} as a sum. Make an array with each element of the sum and use the "cumsum" (`numpy.cumsum`) operation to compute the accumulative sum: `numpy.cumsum([1,3,5])` is `[1,4,9]`.

e) Create a class `MeshCalculus` that can integrate and differentiate mesh functions. The class can just define some methods that call the previously implemented Python functions. Here is an example on the usage:

```
import numpy as np
calc = MeshCalculus(vectorized=True)
x = np.linspace(0, 1, 11)         # mesh
f = np.exp(x)                     # mesh function
df = calc.differentiate(f, x)     # discrete derivative
F = calc.integrate(f, x)          # discrete anti-derivative
```

Filename: `mesh_calculus_1D`.

2.6 Generalization: Reflecting Boundaries

The boundary condition $u = 0$ in a wave equation reflects the wave, but u changes sign at the boundary, while the condition $u_x = 0$ reflects the wave as a mirror and preserves the sign, see a web page[5] or a movie file[6] for demonstration.

Our next task is to explain how to implement the boundary condition $u_x = 0$, which is more complicated to express numerically and also to implement than a given value of u. We shall present two methods for implementing $u_x = 0$ in a finite difference scheme, one based on deriving a modified stencil at the boundary, and another one based on extending the mesh with ghost cells and ghost points.

[5] http://tinyurl.com/hbcasmj/book/html/mov-wave/demo_BC_gaussian/index.html

[6] http://tinyurl.com/gokgkov/mov-wave/demo_BC_gaussian/movie.flv

2.6.1 Neumann Boundary Condition

When a wave hits a boundary and is to be reflected back, one applies the condition

$$\frac{\partial u}{\partial n} \equiv \boldsymbol{n} \cdot \nabla u = 0 . \tag{2.35}$$

The derivative $\partial/\partial n$ is in the outward normal direction from a general boundary. For a 1D domain $[0, L]$, we have that

$$\left.\frac{\partial}{\partial n}\right|_{x=L} = \left.\frac{\partial}{\partial x}\right|_{x=L}, \quad \left.\frac{\partial}{\partial n}\right|_{x=0} = -\left.\frac{\partial}{\partial x}\right|_{x=0} .$$

Boundary condition terminology
Boundary conditions that specify the value of $\partial u/\partial n$ (or shorter u_n) are known as Neumann[7] conditions, while Dirichlet conditions[8] refer to specifications of u. When the values are zero ($\partial u/\partial n = 0$ or $u = 0$) we speak about *homogeneous* Neumann or Dirichlet conditions.

2.6.2 Discretization of Derivatives at the Boundary

How can we incorporate the condition (2.35) in the finite difference scheme? Since we have used central differences in all the other approximations to derivatives in the scheme, it is tempting to implement (2.35) at $x = 0$ and $t = t_n$ by the difference

$$[D_{2x}u]^n_0 = \frac{u^n_{-1} - u^n_1}{2\Delta x} = 0 . \tag{2.36}$$

The problem is that u^n_{-1} is not a u value that is being computed since the point is outside the mesh. However, if we combine (2.36) with the scheme

$$u^{n+1}_i = -u^{n-1}_i + 2u^n_i + C^2 \left(u^n_{i+1} - 2u^n_i + u^n_{i-1}\right), \tag{2.37}$$

for $i = 0$, we can eliminate the fictitious value u^n_{-1}. We see that $u^n_{-1} = u^n_1$ from (2.36), which can be used in (2.37) to arrive at a modified scheme for the boundary point u^{n+1}_0:

$$u^{n+1}_i = -u^{n-1}_i + 2u^n_i + 2C^2 \left(u^n_{i+1} - u^n_i\right), \quad i = 0 . \tag{2.38}$$

Figure 2.4 visualizes this equation for computing u^3_0 in terms of u^2_0, u^1_0, and u^2_1.

Similarly, (2.35) applied at $x = L$ is discretized by a central difference

$$\frac{u^n_{N_x+1} - u^n_{N_x-1}}{2\Delta x} = 0 . \tag{2.39}$$

[7] http://en.wikipedia.org/wiki/Neumann_boundary_condition
[8] http://en.wikipedia.org/wiki/Dirichlet_conditions

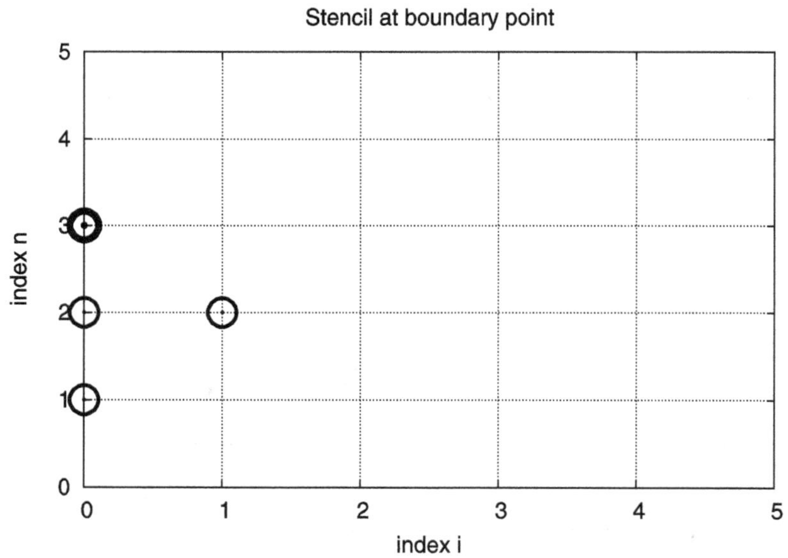

Fig. 2.4 Modified stencil at a boundary with a Neumann condition

Combined with the scheme for $i = N_x$ we get a modified scheme for the boundary value $u_{N_x}^{n+1}$:

$$u_i^{n+1} = -u_i^{n-1} + 2u_i^n + 2C^2\left(u_{i-1}^n - u_i^n\right), \quad i = N_x. \tag{2.40}$$

The modification of the scheme at the boundary is also required for the special formula for the first time step. How the stencil moves through the mesh and is modified at the boundary can be illustrated by an animation in a web page[9] or a movie file[10].

2.6.3 Implementation of Neumann Conditions

We have seen in the preceding section that the special formulas for the boundary points arise from replacing u_{i-1}^n by u_{i+1}^n when computing u_i^{n+1} from the stencil formula for $i = 0$. Similarly, we replace u_{i+1}^n by u_{i-1}^n in the stencil formula for $i = N_x$. This observation can conveniently be used in the coding: we just work with the general stencil formula, but write the code such that it is easy to replace `u[i-1]` by `u[i+1]` and vice versa. This is achieved by having the indices `i+1` and `i-1` as variables `ip1` (`i` plus 1) and `im1` (`i` minus 1), respectively. At the boundary we can easily define `im1=i+1` while we use `im1=i-1` in the internal parts of the mesh. Here are the details of the implementation (note that the updating formula for `u[i]` is the general stencil formula):

[9] http://tinyurl.com/hbcasmj/book/html/mov-wave/N_stencil_gpl/index.html
[10] http://tinyurl.com/gokgkov/mov-wave/N_stencil_gpl/movie.ogg

```
i = 0
ip1 = i+1
im1 = ip1  # i-1 -> i+1
u[i] = u_n[i] + C2*(u_n[im1] - 2*u_n[i] + u_n[ip1])

i = Nx
im1 = i-1
ip1 = im1  # i+1 -> i-1
u[i] = u_n[i] + C2*(u_n[im1] - 2*u_n[i] + u_n[ip1])
```

We can in fact create one loop over both the internal and boundary points and use only one updating formula:

```
for i in range(0, Nx+1):
    ip1 = i+1 if i < Nx else i-1
    im1 = i-1 if i > 0  else i+1
    u[i] = u_n[i] + C2*(u_n[im1] - 2*u_n[i] + u_n[ip1])
```

The program `wave1D_n0.py` contains a complete implementation of the 1D wave equation with boundary conditions $u_x = 0$ at $x = 0$ and $x = L$.

It would be nice to modify the `test_quadratic` test case from the `wave1D_u0.py` with Dirichlet conditions, described in Sect. 2.4.3. However, the Neumann conditions require the polynomial variation in the x direction to be of third degree, which causes challenging problems when designing a test where the numerical solution is known exactly. Exercise 2.15 outlines ideas and code for this purpose. The only test in `wave1D_n0.py` is to start with a plug wave at rest and see that the initial condition is reached again perfectly after one period of motion, but such a test requires $C = 1$ (so the numerical solution coincides with the exact solution of the PDE, see Sect. 2.10.4).

2.6.4 Index Set Notation

To improve our mathematical writing and our implementations, it is wise to introduce a special notation for index sets. This means that we write x_i, followed by $i \in \mathcal{I}_x$, instead of $i = 0, \ldots, N_x$. Obviously, $\mathcal{I}_x$ must be the index set $\mathcal{I}_x = \{0, \ldots, N_x\}$, but it is often advantageous to have a symbol for this set rather than specifying all its elements (all the time, as we have done up to now). This new notation saves writing and makes specifications of algorithms and their implementation as computer code simpler.

The first index in the set will be denoted $\mathcal{I}_x^0$ and the last $\mathcal{I}_x^{-1}$. When we need to skip the first element of the set, we use $\mathcal{I}_x^+$ for the remaining subset $\mathcal{I}_x^+ = \{1, \ldots, N_x\}$. Similarly, if the last element is to be dropped, we write $\mathcal{I}_x^- = \{0, \ldots, N_x - 1\}$ for the remaining indices. All the indices corresponding to inner grid points are specified by $\mathcal{I}_x^i = \{1, \ldots, N_x - 1\}$. For the time domain we find it natural to explicitly use 0 as the first index, so we will usually write $n = 0$ and t_0 rather than $n = \mathcal{I}_t^0$. We also avoid notation like $x_{\mathcal{I}_x^{-1}}$ and will instead use x_i, $i = \mathcal{I}_x^{-1}$.

The Python code associated with index sets applies the following conventions:

Notation	Python
$\mathcal{I}_x$	`Ix`
$\mathcal{I}_x^0$	`Ix[0]`
$\mathcal{I}_x^{-1}$	`Ix[-1]`
$\mathcal{I}_x^-$	`Ix[:-1]`
$\mathcal{I}_x^+$	`Ix[1:]`
$\mathcal{I}_x^i$	`Ix[1:-1]`

Why index sets are useful

An important feature of the index set notation is that it keeps our formulas and code independent of how we count mesh points. For example, the notation $i \in \mathcal{I}_x$ or $i = \mathcal{I}_x^0$ remains the same whether $\mathcal{I}_x$ is defined as above or as starting at 1, i.e., $\mathcal{I}_x = \{1, \ldots, Q\}$. Similarly, we can in the code define `Ix=range(Nx+1)` or `Ix=range(1,Q)`, and expressions like `Ix[0]` and `Ix[1:-1]` remain correct. One application where the index set notation is convenient is conversion of code from a language where arrays has base index 0 (e.g., Python and C) to languages where the base index is 1 (e.g., MATLAB and Fortran). Another important application is implementation of Neumann conditions via ghost points (see next section).

For the current problem setting in the x, t plane, we work with the index sets

$$\mathcal{I}_x = \{0, \ldots, N_x\}, \quad \mathcal{I}_t = \{0, \ldots, N_t\}, \tag{2.41}$$

defined in Python as

```
Ix = range(0, Nx+1)
It = range(0, Nt+1)
```

A finite difference scheme can with the index set notation be specified as

$$\begin{aligned}
u_i^{n+1} &= u_i^n - \frac{1}{2}C^2\left(u_{i+1}^n - 2u_i^n + u_{i-1}^n\right), \quad i \in \mathcal{I}_x^i,\ n = 0,\\
u_i^{n+1} &= -u_i^{n-1} + 2u_i^n + C^2\left(u_{i+1}^n - 2u_i^n + u_{i-1}^n\right), \quad i \in \mathcal{I}_x^i,\ n \in \mathcal{I}_t^i,\\
u_i^{n+1} &= 0, \quad i = \mathcal{I}_x^0,\ n \in \mathcal{I}_t^-,\\
u_i^{n+1} &= 0, \quad i = \mathcal{I}_x^{-1},\ n \in \mathcal{I}_t^-.
\end{aligned}$$

The corresponding implementation becomes

```
# Initial condition
for i in Ix[1:-1]:
    u[i] = u_n[i] - 0.5*C2*(u_n[i-1] - 2*u_n[i] + u_n[i+1])
```

```
# Time loop
for n in It[1:-1]:
    # Compute internal points
    for i in Ix[1:-1]:
        u[i] = - u_nm1[i] + 2*u_n[i] + \
               C2*(u_n[i-1] - 2*u_n[i] + u_n[i+1])
    # Compute boundary conditions
    i = Ix[0];  u[i] = 0
    i = Ix[-1]; u[i] = 0
```

Notice

The program wave1D_dn.py applies the index set notation and solves the 1D wave equation $u_{tt} = c^2 u_{xx} + f(x,t)$ with quite general boundary and initial conditions:

- $x = 0$: $u = U_0(t)$ or $u_x = 0$
- $x = L$: $u = U_L(t)$ or $u_x = 0$
- $t = 0$: $u = I(x)$
- $t = 0$: $u_t = V(x)$

The program combines Dirichlet and Neumann conditions, scalar and vectorized implementation of schemes, and the index set notation into one piece of code. A lot of test examples are also included in the program:

- A rectangular plug-shaped initial condition. (For $C = 1$ the solution will be a rectangle that jumps one cell per time step, making the case well suited for verification.)
- A Gaussian function as initial condition.
- A triangular profile as initial condition, which resembles the typical initial shape of a guitar string.
- A sinusoidal variation of u at $x = 0$ and either $u = 0$ or $u_x = 0$ at $x = L$.
- An analytical solution $u(x,t) = \cos(m\pi t/L)\sin(\frac{1}{2}m\pi x/L)$, which can be used for convergence rate tests.

2.6.5 Verifying the Implementation of Neumann Conditions

How can we test that the Neumann conditions are correctly implemented? The `solver` function in the `wave1D_dn.py` program described in the box above accepts Dirichlet or Neumann conditions at $x = 0$ and $x = L$. It is tempting to apply a quadratic solution as described in Sect. 2.2.1 and 2.3.3, but it turns out that this solution is no longer an exact solution of the discrete equations if a Neumann condition is implemented on the boundary. A linear solution does not help since we only have homogeneous Neumann conditions in `wave1D_dn.py`, and we are consequently left with testing just a constant solution: $u = \text{const}$.

```
def test_constant():
    """
    Check the scalar and vectorized versions for
    a constant u(x,t). We simulate in [0, L] and apply
    Neumann and Dirichlet conditions at both ends.
    """
    u_const = 0.45
    u_exact = lambda x, t: u_const
    I = lambda x: u_exact(x, 0)
    V = lambda x: 0
    f = lambda x, t: 0

    def assert_no_error(u, x, t, n):
        u_e = u_exact(x, t[n])
        diff = np.abs(u - u_e).max()
        msg = 'diff=%E, t_%d=%g' % (diff, n, t[n])
        tol = 1E-13
        assert diff < tol, msg

    for U_0 in (None, lambda t: u_const):
        for U_L in (None, lambda t: u_const):
            L = 2.5
            c = 1.5
            C = 0.75
            Nx = 3  # Very coarse mesh for this exact test
            dt = C*(L/Nx)/c
            T = 18  # long time integration

            solver(I, V, f, c, U_0, U_L, L, dt, C, T,
                   user_action=assert_no_error,
                   version='scalar')
            solver(I, V, f, c, U_0, U_L, L, dt, C, T,
                   user_action=assert_no_error,
                   version='vectorized')
            print U_0, U_L
```

The quadratic solution is very useful for testing, but it requires Dirichlet conditions at both ends.

Another test may utilize the fact that the approximation error vanishes when the Courant number is unity. We can, for example, start with a plug profile as initial condition, let this wave split into two plug waves, one in each direction, and check that the two plug waves come back and form the initial condition again after "one period" of the solution process. Neumann conditions can be applied at both ends. A proper test function reads

```
def test_plug():
    """Check that an initial plug is correct back after one period."""
    L = 1.0
    c = 0.5
    dt = (L/10)/c  # Nx=10
    I = lambda x: 0 if abs(x-L/2.0) > 0.1 else 1
```

```
u_s, x, t, cpu = solver(
    I=I,
    V=None, f=None, c=0.5, U_0=None, U_L=None, L=L,
    dt=dt, C=1, T=4, user_action=None, version='scalar')
u_v, x, t, cpu = solver(
    I=I,
    V=None, f=None, c=0.5, U_0=None, U_L=None, L=L,
    dt=dt, C=1, T=4, user_action=None, version='vectorized')
tol = 1E-13
diff = abs(u_s - u_v).max()
assert diff < tol
u_0 = np.array([I(x_) for x_ in x])
diff = np.abs(u_s - u_0).max()
assert diff < tol
```

Other tests must rely on an unknown approximation error, so effectively we are left with tests on the convergence rate.

2.6.6 Alternative Implementation via Ghost Cells

Idea Instead of modifying the scheme at the boundary, we can introduce extra points outside the domain such that the fictitious values u^n_{-1} and $u^n_{N_x+1}$ are defined in the mesh. Adding the intervals $[-\Delta x, 0]$ and $[L, L + \Delta x]$, known as *ghost cells*, to the mesh gives us all the needed mesh points, corresponding to $= -1, 0, \ldots, N_x, N_x + 1$. The extra points with $i = -1$ and $i = N_x + 1$ are known as *ghost points*, and values at these points, u^n_{-1} and $u^n_{N_x+1}$, are called *ghost values*.

The important idea is to ensure that we always have

$$u^n_{-1} = u^n_1 \quad \text{and} \quad u^n_{N_x+1} = u^n_{N_x-1},$$

because then the application of the standard scheme at a boundary point $i = 0$ or $i = N_x$ will be correct and guarantee that the solution is compatible with the boundary condition $u_x = 0$.

Some readers may find it strange to just extend the domain with ghost cells as a general technique, because in some problems there is a completely different medium with different physics and equations right outside of a boundary. Nevertheless, one should view the ghost cell technique as a purely mathematical technique, which is valid in the limit $\Delta x \to 0$ and helps us to implement derivatives.

Implementation The u array now needs extra elements corresponding to the ghost points. Two new point values are needed:

```
u   = zeros(Nx+3)
```

The arrays `u_n` and `u_nm1` must be defined accordingly.

Unfortunately, a major indexing problem arises with ghost cells. The reason is that Python indices *must* start at 0 and `u[-1]` will always mean the last element in `u`. This fact gives, apparently, a mismatch between the mathematical indices $i = -1, 0, \ldots, N_x + 1$ and the Python indices running over `u`: `0,..,Nx+2`. One remedy is to change the mathematical indexing of i in the scheme and write

$$u_i^{n+1} = \cdots, \quad i = 1, \ldots, N_x + 1,$$

instead of $i = 0, \ldots, N_x$ as we have previously used. The ghost points now correspond to $i = 0$ and $i = N_x + 1$. A better solution is to use the ideas of Sect. 2.6.4: we hide the specific index value in an index set and operate with inner and boundary points using the index set notation.

To this end, we define `u` with proper length and `Ix` to be the corresponding indices for the real physical mesh points $(1, 2, \ldots, N_x + 1)$:

```
u = zeros(Nx+3)
Ix = range(1, u.shape[0]-1)
```

That is, the boundary points have indices `Ix[0]` and `Ix[-1]` (as before). We first update the solution at all physical mesh points (i.e., interior points in the mesh):

```
for i in Ix:
    u[i] = - u_nm1[i] + 2*u_n[i] + \
             C2*(u_n[i-1] - 2*u_n[i] + u_n[i+1])
```

The indexing becomes a bit more complicated when we call functions like `V(x)` and `f(x, t)`, as we must remember that the appropriate x coordinate is given as `x[i-Ix[0]]`:

```
for i in Ix:
    u[i] = u_n[i] + dt*V(x[i-Ix[0]]) + \
             0.5*C2*(u_n[i-1] - 2*u_n[i] + u_n[i+1]) + \
             0.5*dt2*f(x[i-Ix[0]], t[0])
```

It remains to update the solution at ghost points, i.e., `u[0]` and `u[-1]` (or `u[Nx+2]`). For a boundary condition $u_x = 0$, the ghost value must equal the value at the associated inner mesh point. Computer code makes this statement precise:

```
i = Ix[0]              # x=0 boundary
u[i-1] = u[i+1]
i = Ix[-1]             # x=L boundary
u[i+1] = u[i-1]
```

The physical solution to be plotted is now in `u[1:-1]`, or equivalently `u[Ix[0]:Ix[-1]+1]`, so this slice is the quantity to be returned from a solver function. A complete implementation appears in the program `wave1D_n0_ghost.py`.

Warning

We have to be careful with how the spatial and temporal mesh points are stored. Say we let x be the physical mesh points,

```
x = linspace(0, L, Nx+1)
```

"Standard coding" of the initial condition,

```
for i in Ix:
    u_n[i] = I(x[i])
```

becomes wrong, since `u_n` and `x` have different lengths and the index `i` corresponds to two different mesh points. In fact, `x[i]` corresponds to `u[1+i]`. A correct implementation is

```
for i in Ix:
    u_n[i] = I(x[i-Ix[0]])
```

Similarly, a source term usually coded as `f(x[i], t[n])` is incorrect if `x` is defined to be the physical points, so `x[i]` must be replaced by `x[i-Ix[0]]`.

An alternative remedy is to let `x` also cover the ghost points such that `u[i]` is the value at `x[i]`.

The ghost cell is only added to the boundary where we have a Neumann condition. Suppose we have a Dirichlet condition at $x = L$ and a homogeneous Neumann condition at $x = 0$. One ghost cell $[-\Delta x, 0]$ is added to the mesh, so the index set for the physical points becomes $\{1, \ldots, N_x + 1\}$. A relevant implementation is

```
u = zeros(Nx+2)
Ix = range(1, u.shape[0])
...
for i in Ix[:-1]:
    u[i] = - u_nm1[i] + 2*u_n[i] + \
           C2*(u_n[i-1] - 2*u_n[i] + u_n[i+1]) + \
           dt2*f(x[i-Ix[0]], t[n])
i = Ix[-1]
u[i] = U_0        # set Dirichlet value
i = Ix[0]
u[i-1] = u[i+1]   # update ghost value
```

The physical solution to be plotted is now in `u[1:]` or (as always) `u[Ix[0]:Ix[-1]+1]`.

2.7 Generalization: Variable Wave Velocity

Our next generalization of the 1D wave equation (2.1) or (2.17) is to allow for a variable wave velocity c: $c = c(x)$, usually motivated by wave motion in a domain composed of different physical media. When the media differ in physical properties like density or porosity, the wave velocity c is affected and will depend on the position in space. Figure 2.5 shows a wave propagating in one medium $[0, 0.7] \cup [0.9, 1]$ with wave velocity c_1 (left) before it enters a second medium $(0.7, 0.9)$ with wave velocity c_2 (right). When the wave meets the boundary where c jumps from c_1 to c_2, a part of the wave is reflected back into the first medium (the *reflected* wave), while one part is transmitted through the second medium (the *transmitted* wave).

2.7.1 The Model PDE with a Variable Coefficient

Instead of working with the squared quantity $c^2(x)$, we shall for notational convenience introduce $q(x) = c^2(x)$. A 1D wave equation with variable wave velocity often takes the form

$$\frac{\partial^2 u}{\partial t^2} = \frac{\partial}{\partial x}\left(q(x)\frac{\partial u}{\partial x}\right) + f(x,t). \qquad (2.42)$$

This is the most frequent form of a wave equation with variable wave velocity, but other forms also appear, see Sect. 2.14.1 and equation (2.125).

As usual, we sample (2.42) at a mesh point,

$$\frac{\partial^2}{\partial t^2}u(x_i,t_n) = \frac{\partial}{\partial x}\left(q(x_i)\frac{\partial}{\partial x}u(x_i,t_n)\right) + f(x_i,t_n),$$

where the only new term to discretize is

$$\frac{\partial}{\partial x}\left(q(x_i)\frac{\partial}{\partial x}u(x_i,t_n)\right) = \left[\frac{\partial}{\partial x}\left(q(x)\frac{\partial u}{\partial x}\right)\right]_i^n.$$

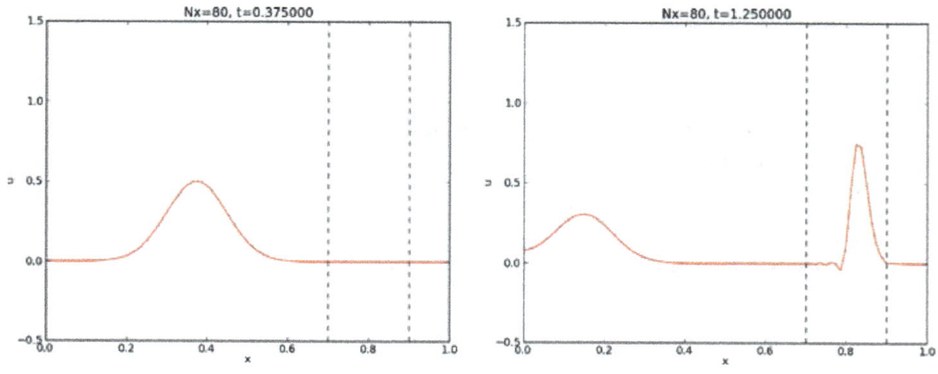

Fig. 2.5 *Left:* wave entering another medium; *right:* transmitted and reflected wave

2.7.2 Discretizing the Variable Coefficient

The principal idea is to first discretize the outer derivative. Define

$$\phi = q(x)\frac{\partial u}{\partial x},$$

and use a centered derivative around $x = x_i$ for the derivative of ϕ:

$$\left[\frac{\partial \phi}{\partial x}\right]_i^n \approx \frac{\phi_{i+\frac{1}{2}} - \phi_{i-\frac{1}{2}}}{\Delta x} = [D_x\phi]_i^n .$$

Then discretize

$$\phi_{i+\frac{1}{2}} = q_{i+\frac{1}{2}}\left[\frac{\partial u}{\partial x}\right]_{i+\frac{1}{2}}^n \approx q_{i+\frac{1}{2}}\frac{u_{i+1}^n - u_i^n}{\Delta x} = [qD_xu]_{i+\frac{1}{2}}^n .$$

Similarly,

$$\phi_{i-\frac{1}{2}} = q_{i-\frac{1}{2}}\left[\frac{\partial u}{\partial x}\right]_{i-\frac{1}{2}}^n \approx q_{i-\frac{1}{2}}\frac{u_i^n - u_{i-1}^n}{\Delta x} = [qD_xu]_{i-\frac{1}{2}}^n .$$

These intermediate results are now combined to

$$\left[\frac{\partial}{\partial x}\left(q(x)\frac{\partial u}{\partial x}\right)\right]_i^n \approx \frac{1}{\Delta x^2}\left(q_{i+\frac{1}{2}}\left(u_{i+1}^n - u_i^n\right) - q_{i-\frac{1}{2}}\left(u_i^n - u_{i-1}^n\right)\right) . \tag{2.43}$$

With operator notation we can write the discretization as

$$\left[\frac{\partial}{\partial x}\left(q(x)\frac{\partial u}{\partial x}\right)\right]_i^n \approx [D_x(\overline{q}^x D_xu)]_i^n . \tag{2.44}$$

Do not use the chain rule on the spatial derivative term!

Many are tempted to use the chain rule on the term $\frac{\partial}{\partial x}\left(q(x)\frac{\partial u}{\partial x}\right)$, but this is not a good idea when discretizing such a term.

The term with a variable coefficient expresses the net flux qu_x into a small volume (i.e., interval in 1D):

$$\frac{\partial}{\partial x}\left(q(x)\frac{\partial u}{\partial x}\right) \approx \frac{1}{\Delta x}(q(x+\Delta x)u_x(x+\Delta x) - q(x)u_x(x)) .$$

Our discretization reflects this principle directly: qu_x at the right end of the cell minus qu_x at the left end, because this follows from the formula (2.43) or $[D_x(qD_xu)]_i^n$.

When using the chain rule, we get two terms $qu_{xx} + q_xu_x$. The typical discretization is

$$[D_xqD_xu + D_{2x}qD_{2x}u]_i^n , \tag{2.45}$$

Writing this out shows that it is different from $[D_x(qD_xu)]_i^n$ and lacks the physical interpretation of net flux into a cell. With a smooth and slowly varying $q(x)$

the differences between the two discretizations are not substantial. However, when q exhibits (potentially large) jumps, $[D_x(qD_xu)]^n_i$ with harmonic averaging of q yields a better solution than arithmetic averaging or (2.45). In the literature, the discretization $[D_x(qD_xu)]^n_i$ totally dominates and very few mention the alternative in (2.45).

2.7.3 Computing the Coefficient Between Mesh Points

If q is a known function of x, we can easily evaluate $q_{i+\frac{1}{2}}$ simply as $q(x_{i+\frac{1}{2}})$ with $x_{i+\frac{1}{2}} = x_i + \frac{1}{2}\Delta x$. However, in many cases c, and hence q, is only known as a discrete function, often at the mesh points x_i. Evaluating q between two mesh points x_i and x_{i+1} must then be done by *interpolation* techniques, of which three are of particular interest in this context:

$$q_{i+\frac{1}{2}} \approx \frac{1}{2}(q_i + q_{i+1}) = [\overline{q}^x]_i \qquad \text{(arithmetic mean)} \tag{2.46}$$

$$q_{i+\frac{1}{2}} \approx 2\left(\frac{1}{q_i} + \frac{1}{q_{i+1}}\right)^{-1} \qquad \text{(harmonic mean)} \tag{2.47}$$

$$q_{i+\frac{1}{2}} \approx (q_i q_{i+1})^{1/2} \qquad \text{(geometric mean)} \tag{2.48}$$

The arithmetic mean in (2.46) is by far the most commonly used averaging technique and is well suited for smooth $q(x)$ functions. The harmonic mean is often preferred when $q(x)$ exhibits large jumps (which is typical for geological media). The geometric mean is less used, but popular in discretizations to linearize quadratic nonlinearities (see Sect. 1.10.2 for an example).

With the operator notation from (2.46) we can specify the discretization of the complete variable-coefficient wave equation in a compact way:

$$[D_tD_tu = D_x\overline{q}^x D_xu + f]^n_i \,. \tag{2.49}$$

Strictly speaking, $[D_x\overline{q}^x D_xu]^n_i = [D_x(\overline{q}^x D_xu)]^n_i$.

From the compact difference notation we immediately see what kind of differences that each term is approximated with. The notation $\overline{q}^x$ also specifies that the variable coefficient is approximated by an arithmetic mean, the definition being $[\overline{q}^x]_{i+\frac{1}{2}} = (q_i + q_{i+1})/2$.

Before implementing, it remains to solve (2.49) with respect to u^{n+1}_i:

$$\begin{aligned} u^{n+1}_i &= -u^{n-1}_i + 2u^n_i \\ &\quad + \left(\frac{\Delta t}{\Delta x}\right)^2 \left(\frac{1}{2}(q_i + q_{i+1})(u^n_{i+1} - u^n_i) - \frac{1}{2}(q_i + q_{i-1})(u^n_i - u^n_{i-1})\right) \\ &\quad + \Delta t^2 f^n_i \,. \end{aligned} \tag{2.50}$$

2.7.4 How a Variable Coefficient Affects the Stability

The stability criterion derived later (Sect. 2.10.3) reads $\Delta t \leq \Delta x/c$. If $c = c(x)$, the criterion will depend on the spatial location. We must therefore choose a Δt that is small enough such that no mesh cell has $\Delta t > \Delta x/c(x)$. That is, we must use the largest c value in the criterion:

$$\Delta t \leq \beta \frac{\Delta x}{\max_{x\in[0,L]} c(x)} . \tag{2.51}$$

The parameter β is included as a safety factor: in some problems with a significantly varying c it turns out that one must choose $\beta < 1$ to have stable solutions ($\beta = 0.9$ may act as an all-round value).

A different strategy to handle the stability criterion with variable wave velocity is to use a spatially varying Δt. While the idea is mathematically attractive at first sight, the implementation quickly becomes very complicated, so we stick to a constant Δt and a worst case value of $c(x)$ (with a safety factor β).

2.7.5 Neumann Condition and a Variable Coefficient

Consider a Neumann condition $\partial u/\partial x = 0$ at $x = L = N_x\Delta x$, discretized as

$$[D_{2x}u]^n_i = \frac{u^n_{i+1} - u^n_{i-1}}{2\Delta x} = 0 \quad \Rightarrow \quad u^n_{i+1} = u^n_{i-1},$$

for $i = N_x$. Using the scheme (2.50) at the end point $i = N_x$ with $u^n_{i+1} = u^n_{i-1}$ results in

$$\begin{aligned} u^{n+1}_i &= -u^{n-1}_i + 2u^n_i \\ &\quad + \left(\frac{\Delta t}{\Delta x}\right)^2 \left(q_{i+\frac{1}{2}}(u^n_{i-1} - u^n_i) - q_{i-\frac{1}{2}}(u^n_i - u^n_{i-1})\right) + \Delta t^2 f^n_i & (2.52) \\ &= -u^{n-1}_i + 2u^n_i + \left(\frac{\Delta t}{\Delta x}\right)^2 (q_{i+\frac{1}{2}} + q_{i-\frac{1}{2}})(u^n_{i-1} - u^n_i) + \Delta t^2 f^n_i & (2.53) \\ &\approx -u^{n-1}_i + 2u^n_i + \left(\frac{\Delta t}{\Delta x}\right)^2 2q_i(u^n_{i-1} - u^n_i) + \Delta t^2 f^n_i . & (2.54) \end{aligned}$$

Here we used the approximation

$$\begin{aligned} q_{i+\frac{1}{2}} + q_{i-\frac{1}{2}} &= q_i + \left(\frac{dq}{dx}\right)_i \Delta x + \left(\frac{d^2q}{dx^2}\right)_i \Delta x^2 + \cdots \\ &\quad + q_i - \left(\frac{dq}{dx}\right)_i \Delta x + \left(\frac{d^2q}{dx^2}\right)_i \Delta x^2 + \cdots \\ &= 2q_i + 2\left(\frac{d^2q}{dx^2}\right)_i \Delta x^2 + \mathcal{O}(\Delta x^4) \\ &\approx 2q_i . \end{aligned} \tag{2.55}$$

An alternative derivation may apply the arithmetic mean of $q_{n-\frac{1}{2}}$ and $q_{n+\frac{1}{2}}$ in (2.53), leading to the term

$$\left(q_i + \frac{1}{2}(q_{i+1} + q_{i-1})\right)(u_{i-1}^n - u_i^n)\,.$$

Since $\frac{1}{2}(q_{i+1} + q_{i-1}) = q_i + \mathcal{O}(\Delta x^2)$, we can approximate with $2q_i(u_{i-1}^n - u_i^n)$ for $i = N_x$ and get the same term as we did above.

A common technique when implementing $\partial u/\partial x = 0$ boundary conditions, is to assume $dq/dx = 0$ as well. This implies $q_{i+1} = q_{i-1}$ and $q_{i+1/2} = q_{i-1/2}$ for $i = N_x$. The implications for the scheme are

$$\begin{aligned} u_i^{n+1} &= -u_i^{n-1} + 2u_i^n \\ &\quad + \left(\frac{\Delta t}{\Delta x}\right)^2 \left(q_{i+\frac{1}{2}}(u_{i-1}^n - u_i^n) - q_{i-\frac{1}{2}}(u_i^n - u_{i-1}^n)\right) \\ &\quad + \Delta t^2 f_i^n & (2.56) \\ &= -u_i^{n-1} + 2u_i^n + \left(\frac{\Delta t}{\Delta x}\right)^2 2q_{i-\frac{1}{2}}(u_{i-1}^n - u_i^n) + \Delta t^2 f_i^n\,. & (2.57) \end{aligned}$$

2.7.6 Implementation of Variable Coefficients

The implementation of the scheme with a variable wave velocity $q(x) = c^2(x)$ may assume that q is available as an array `q[i]` at the spatial mesh points. The following loop is a straightforward implementation of the scheme (2.50):

```
for i in range(1, Nx):
    u[i] = - u_nm1[i] + 2*u_n[i] + \
           C2*(0.5*(q[i] + q[i+1])*(u_n[i+1] - u_n[i])  - \
               0.5*(q[i] + q[i-1])*(u_n[i] - u_n[i-1])) + \
           dt2*f(x[i], t[n])
```

The coefficient `C2` is now defined as `(dt/dx)**2`, i.e., *not* as the squared Courant number, since the wave velocity is variable and appears inside the parenthesis.

With Neumann conditions $u_x = 0$ at the boundary, we need to combine this scheme with the discrete version of the boundary condition, as shown in Sect. 2.7.5. Nevertheless, it would be convenient to reuse the formula for the interior points and just modify the indices `ip1=i+1` and `im1=i-1` as we did in Sect. 2.6.3. Assuming $dq/dx = 0$ at the boundaries, we can implement the scheme at the boundary with the following code.

```
i = 0
ip1 = i+1
im1 = ip1
u[i] = - u_nm1[i] + 2*u_n[i] + \
       C2*(0.5*(q[i] + q[ip1])*(u_n[ip1] - u_n[i])  - \
           0.5*(q[i] + q[im1])*(u_n[i] - u_n[im1])) + \
       dt2*f(x[i], t[n])
```

With ghost cells we can just reuse the formula for the interior points also at the boundary, provided that the ghost values of both u and q are correctly updated to ensure $u_x = 0$ and $q_x = 0$.

A vectorized version of the scheme with a variable coefficient at internal mesh points becomes

```
u[1:-1] = - u_nm1[1:-1] + 2*u_n[1:-1] + \
          C2*(0.5*(q[1:-1] + q[2:])*(u_n[2:] - u_n[1:-1]) -
              0.5*(q[1:-1] + q[:-2])*(u_n[1:-1] - u_n[:-2])) + \
          dt2*f(x[1:-1], t[n])
```

2.7.7 A More General PDE Model with Variable Coefficients

Sometimes a wave PDE has a variable coefficient in front of the time-derivative term:

$$\varrho(x)\frac{\partial^2 u}{\partial t^2} = \frac{\partial}{\partial x}\left(q(x)\frac{\partial u}{\partial x}\right) + f(x,t)\,. \tag{2.58}$$

One example appears when modeling elastic waves in a rod with varying density, cf. (2.14.1) with $\varrho(x)$.

A natural scheme for (2.58) is

$$[\varrho D_t D_t u = D_x \overline{q}^x D_x u + f]_i^n\,. \tag{2.59}$$

We realize that the ϱ coefficient poses no particular difficulty, since ϱ enters the formula just as a simple factor in front of a derivative. There is hence no need for any averaging of ϱ. Often, ϱ will be moved to the right-hand side, also without any difficulty:

$$[D_t D_t u = \varrho^{-1} D_x \overline{q}^x D_x u + f]_i^n\,. \tag{2.60}$$

2.7.8 Generalization: Damping

Waves die out by two mechanisms. In 2D and 3D the energy of the wave spreads out in space, and energy conservation then requires the amplitude to decrease. This effect is not present in 1D. Damping is another cause of amplitude reduction. For example, the vibrations of a string die out because of damping due to air resistance and non-elastic effects in the string.

The simplest way of including damping is to add a first-order derivative to the equation (in the same way as friction forces enter a vibrating mechanical system):

$$\frac{\partial^2 u}{\partial t^2} + b\frac{\partial u}{\partial t} = c^2\frac{\partial^2 u}{\partial x^2} + f(x,t), \tag{2.61}$$

where $b \geq 0$ is a prescribed damping coefficient.

A typical discretization of (2.61) in terms of centered differences reads

$$[D_t D_t u + bD_{2t}u = c^2 D_x D_x u + f]_i^n\,. \tag{2.62}$$

Writing out the equation and solving for the unknown u_i^{n+1} gives the scheme

$$u_i^{n+1} = \left(1 + \frac{1}{2}b\Delta t\right)^{-1} \left(\left(\frac{1}{2}b\Delta t - 1\right) u_i^{n-1} + 2u_i^n + C^2 \left(u_{i+1}^n - 2u_i^n + u_{i-1}^n\right) + \Delta t^2 f_i^n \right), \tag{2.63}$$

for $i \in \mathcal{I}_x^i$ and $n \geq 1$. New equations must be derived for u_i^1, and for boundary points in case of Neumann conditions.

The damping is very small in many wave phenomena and thus only evident for very long time simulations. This makes the standard wave equation without damping relevant for a lot of applications.

2.8 Building a General 1D Wave Equation Solver

The program `wave1D_dn_vc.py` is a fairly general code for 1D wave propagation problems that targets the following initial-boundary value problem

$$u_{tt} = (c^2(x)u_x)_x + f(x,t), \quad x \in (0,L),\ t \in (0,T] \tag{2.64}$$

$$u(x,0) = I(x), \quad x \in [0,L] \tag{2.65}$$

$$u_t(x,0) = V(t), \quad x \in [0,L] \tag{2.66}$$

$$u(0,t) = U_0(t) \quad \text{or} \quad u_x(0,t) = 0, \quad t \in (0,T] \tag{2.67}$$

$$u(L,t) = U_L(t) \quad \text{or} \quad u_x(L,t) = 0, \quad t \in (0,T]. \tag{2.68}$$

The only new feature here is the time-dependent Dirichlet conditions, but they are trivial to implement:

```
i = Ix[0]  # x=0
u[i] = U_0(t[n+1])

i = Ix[-1] # x=L
u[i] = U_L(t[n+1])
```

The `solver` function is a natural extension of the simplest `solver` function in the initial `wave1D_u0.py` program, extended with Neumann boundary conditions ($u_x = 0$), time-varying Dirichlet conditions, as well as a variable wave velocity. The different code segments needed to make these extensions have been shown and commented upon in the preceding text. We refer to the `solver` function in the `wave1D_dn_vc.py` file for all the details. Note in that `solver` function, however, that the technique of "hashing" is used to check whether a certain simulation has been run before, or not. This technique is further explained in Sect. C.2.3.

The vectorization is only applied inside the time loop, not for the initial condition or the first time steps, since this initial work is negligible for long time simulations in 1D problems.

The following sections explain various more advanced programming techniques applied in the general 1D wave equation solver.

2.8.1 User Action Function as a Class

A useful feature in the `wave1D_dn_vc.py` program is the specification of the `user_action` function as a class. This part of the program may need some motivation and explanation. Although the `plot_u_st` function (and the `PlotMatplotlib` class) in the `wave1D_u0.viz` function remembers the local variables in the `viz` function, it is a cleaner solution to store the needed variables together with the function, which is exactly what a class offers.

The code A class for flexible plotting, cleaning up files, making movie files, like the function `wave1D_u0.viz` did, can be coded as follows:

```
class PlotAndStoreSolution:
    """
    Class for the user_action function in solver.
    Visualizes the solution only.
    """
    def __init__(
        self,
        casename='tmp',       # Prefix in filenames
        umin=-1, umax=1,      # Fixed range of y axis
        pause_between_frames=None,  # Movie speed
        backend='matplotlib',       # or 'gnuplot' or None
        screen_movie=True, # Show movie on screen?
        title='',             # Extra message in title
        skip_frame=1,         # Skip every skip_frame frame
        filename=None):       # Name of file with solutions
        self.casename = casename
        self.yaxis = [umin, umax]
        self.pause = pause_between_frames
        self.backend = backend
        if backend is None:
            # Use native matplotlib
            import matplotlib.pyplot as plt
        elif backend in ('matplotlib', 'gnuplot'):
            module = 'scitools.easyviz.' + backend + '_'
            exec('import %s as plt' % module)
        self.plt = plt
        self.screen_movie = screen_movie
        self.title = title
        self.skip_frame = skip_frame
        self.filename = filename
        if filename is not None:
            # Store time points when u is written to file
            self.t = []
            filenames = glob.glob('.' + self.filename + '*.dat.npz')
            for filename in filenames:
                os.remove(filename)

        # Clean up old movie frames
        for filename in glob.glob('frame_*.png'):
            os.remove(filename)
```

```
def __call__(self, u, x, t, n):
    """
    Callback function user_action, call by solver:
    Store solution, plot on screen and save to file.
    """
    # Save solution u to a file using numpy.savez
    if self.filename is not None:
        name = 'u%04d' % n  # array name
        kwargs = {name: u}
        fname = '.' + self.filename + '_' + name + '.dat'
        np.savez(fname, **kwargs)
        self.t.append(t[n])  # store corresponding time value
        if n == 0:           # save x once
            np.savez('.' + self.filename + '_x.dat', x=x)

    # Animate
    if n % self.skip_frame != 0:
        return
    title = 't=%.3f' % t[n]
    if self.title:
        title = self.title + ' ' + title
    if self.backend is None:
        # native matplotlib animation
        if n == 0:
            self.plt.ion()
            self.lines = self.plt.plot(x, u, 'r-')
            self.plt.axis([x[0], x[-1],
                           self.yaxis[0], self.yaxis[1]])
            self.plt.xlabel('x')
            self.plt.ylabel('u')
            self.plt.title(title)
            self.plt.legend(['t=%.3f' % t[n]])
        else:
            # Update new solution
            self.lines[0].set_ydata(u)
            self.plt.legend(['t=%.3f' % t[n]])
            self.plt.draw()
    else:
        # scitools.easyviz animation
        self.plt.plot(x, u, 'r-',
                      xlabel='x', ylabel='u',
                      axis=[x[0], x[-1],
                            self.yaxis[0], self.yaxis[1]],
                      title=title,
                      show=self.screen_movie)
    # pause
    if t[n] == 0:
        time.sleep(2)  # let initial condition stay 2 s
    else:
        if self.pause is None:
            pause = 0.2 if u.size < 100 else 0
        time.sleep(pause)

    self.plt.savefig('frame_%04d.png' % (n))
```

Dissection Understanding this class requires quite some familiarity with Python in general and class programming in particular. The class supports plotting with Matplotlib (`backend=None`) or SciTools (`backend=matplotlib` or `backend=gnuplot`) for maximum flexibility.

The constructor shows how we can flexibly import the plotting engine as (typically) `scitools.easyviz.gnuplot_` or `scitools.easyviz.matplotlib_` (note the trailing underscore - it is required). With the `screen_movie` parameter we can suppress displaying each movie frame on the screen. Alternatively, for slow movies associated with fine meshes, one can set `skip_frame=10`, causing every 10 frames to be shown.

The `__call__` method makes `PlotAndStoreSolution` instances behave like functions, so we can just pass an instance, say `p`, as the `user_action` argument in the `solver` function, and any call to `user_action` will be a call to `p.__call__`. The `__call__` method plots the solution on the screen, saves the plot to file, and stores the solution in a file for later retrieval.

More details on storing the solution in files appear in Sect. C.2.

2.8.2 Pulse Propagation in Two Media

The function `pulse` in `wave1D_dn_vc.py` demonstrates wave motion in heterogeneous media where c varies. One can specify an interval where the wave velocity is decreased by a factor `slowness_factor` (or increased by making this factor less than one). Figure 2.5 shows a typical simulation scenario.

Four types of initial conditions are available:

1. a rectangular pulse (`plug`),
2. a Gaussian function (`gaussian`),
3. a "cosine hat" consisting of one period of the cosine function (`cosinehat`),
4. half a period of a "cosine hat" (`half-cosinehat`)

These peak-shaped initial conditions can be placed in the middle (`loc='center'`) or at the left end (`loc='left'`) of the domain. With the pulse in the middle, it splits in two parts, each with half the initial amplitude, traveling in opposite directions. With the pulse at the left end, centered at $x = 0$, and using the symmetry condition $\partial u/\partial x = 0$, only a right-going pulse is generated. There is also a left-going pulse, but it travels from $x = 0$ in negative x direction and is not visible in the domain $[0, L]$.

The `pulse` function is a flexible tool for playing around with various wave shapes and jumps in the wave velocity (i.e., discontinuous media). The code is shown to demonstrate how easy it is to reach this flexibility with the building blocks we have already developed:

```
def pulse(
    C=1,            # Maximum Courant number
    Nx=200,         # spatial resolution
    animate=True,
    version='vectorized',
    T=2,            # end time
    loc='left',     # location of initial condition
    pulse_tp='gaussian',  # pulse/init.cond. type
    slowness_factor=2, # inverse of wave vel. in right medium
    medium=[0.7, 0.9], # interval for right medium
    skip_frame=1,      # skip frames in animations
    sigma=0.05         # width measure of the pulse
    ):
    """
    Various peaked-shaped initial conditions on [0,1].
    Wave velocity is decreased by the slowness_factor inside
    medium. The loc parameter can be 'center' or 'left',
    depending on where the initial pulse is to be located.
    The sigma parameter governs the width of the pulse.
    """
    # Use scaled parameters: L=1 for domain length, c_0=1
    # for wave velocity outside the domain.
    L = 1.0
    c_0 = 1.0
    if loc == 'center':
        xc = L/2
    elif loc == 'left':
        xc = 0

    if pulse_tp in ('gaussian','Gaussian'):
        def I(x):
            return np.exp(-0.5*((x-xc)/sigma)**2)
    elif pulse_tp == 'plug':
        def I(x):
            return 0 if abs(x-xc) > sigma else 1
    elif pulse_tp == 'cosinehat':
        def I(x):
            # One period of a cosine
            w = 2
            a = w*sigma
            return 0.5*(1 + np.cos(np.pi*(x-xc)/a)) \
                   if xc - a <= x <= xc + a else 0

    elif pulse_tp == 'half-cosinehat':
        def I(x):
            # Half a period of a cosine
            w = 4
            a = w*sigma
            return np.cos(np.pi*(x-xc)/a) \
                   if xc - 0.5*a <= x <= xc + 0.5*a else 0
    else:
        raise ValueError('Wrong pulse_tp="%s"' % pulse_tp)

    def c(x):
        return c_0/slowness_factor \
               if medium[0] <= x <= medium[1] else c_0
```

```
    umin=-0.5; umax=1.5*I(xc)
    casename = '%s_Nx%s_sf%s' % \
               (pulse_tp, Nx, slowness_factor)
    action = PlotMediumAndSolution(
        medium, casename=casename, umin=umin, umax=umax,
        skip_frame=skip_frame, screen_movie=animate,
        backend=None, filename='tmpdata')

    # Choose the stability limit with given Nx, worst case c
    # (lower C will then use this dt, but smaller Nx)
    dt = (L/Nx)/c_0
    cpu, hashed_input = solver(
        I=I, V=None, f=None, c=c,
        U_0=None, U_L=None,
        L=L, dt=dt, C=C, T=T,
        user_action=action,
        version=version,
        stability_safety_factor=1)

    if cpu > 0:   # did we generate new data?
        action.close_file(hashed_input)
        action.make_movie_file()
    print 'cpu (-1 means no new data generated):', cpu

def convergence_rates(
    u_exact,
    I, V, f, c, U_0, U_L, L,
    dt0, num_meshes,
    C, T, version='scalar',
    stability_safety_factor=1.0):
    """
    Half the time step and estimate convergence rates for
    for num_meshes simulations.
    """
    class ComputeError:
        def __init__(self, norm_type):
            self.error = 0

        def __call__(self, u, x, t, n):
            """Store norm of the error in self.E."""
            error = np.abs(u - u_exact(x, t[n])).max()
            self.error = max(self.error, error)

    E = []
    h = []  # dt, solver adjusts dx such that C=dt*c/dx
    dt = dt0
    for i in range(num_meshes):
        error_calculator = ComputeError('Linf')
        solver(I, V, f, c, U_0, U_L, L, dt, C, T,
               user_action=error_calculator,
               version='scalar',
               stability_safety_factor=1.0)
        E.append(error_calculator.error)
        h.append(dt)
        dt /= 2  # halve the time step for next simulation
    print 'E:', E
    print 'h:', h
    r = [np.log(E[i]/E[i-1])/np.log(h[i]/h[i-1])
         for i in range(1,num_meshes)]
    return r
```

```
def test_convrate_sincos():
    n = m = 2
    L = 1.0
    u_exact = lambda x, t: np.cos(m*np.pi/L*t)*np.sin(m*np.pi/L*x)

    r = convergence_rates(
        u_exact=u_exact,
        I=lambda x: u_exact(x, 0),
        V=lambda x: 0,
        f=0,
        c=1,
        U_0=0,
        U_L=0,
        L=L,
        dt0=0.1,
        num_meshes=6,
        C=0.9,
        T=1,
        version='scalar',
        stability_safety_factor=1.0)
    print 'rates sin(x)*cos(t) solution:', \
          [round(r_,2) for r_ in r]
    assert abs(r[-1] - 2) < 0.002
```

The `PlotMediumAndSolution` class used here is a subclass of `PlotAndStoreSolution` where the medium with reduced c value, as specified by the `medium` interval, is visualized in the plots.

Comment on the choices of discretization parameters

The argument N_x in the `pulse` function does not correspond to the actual spatial resolution of $C < 1$, since the `solver` function takes a fixed Δt and C, and adjusts Δx accordingly. As seen in the `pulse` function, the specified Δt is chosen according to the limit $C = 1$, so if $C < 1$, Δt remains the same, but the `solver` function operates with a larger Δx and smaller N_x than was specified in the call to `pulse`. The practical reason is that we always want to keep Δt fixed such that plot frames and movies are synchronized in time regardless of the value of C (i.e., Δx is varied when the Courant number varies).

The reader is encouraged to play around with the `pulse` function:

```
>>> import wave1D_dn_vc as w
>>> w.pulse(Nx=50, loc='left', pulse_tp='cosinehat', slowness_factor=2)
```

To easily kill the graphics by Ctrl-C and restart a new simulation it might be easier to run the above two statements from the command line with

Terminal

```
Terminal> python -c 'import wave1D_dn_vc as w; w.pulse(...)'
```

2.9 Exercises

Exercise 2.7: Find the analytical solution to a damped wave equation
Consider the wave equation with damping (2.61). The goal is to find an exact solution to a wave problem with damping and zero source term. A starting point is the standing wave solution from Exercise 2.1. It becomes necessary to include a damping term $e^{-\beta t}$ and also have both a sine and cosine component in time:

$$u_{\mathrm{e}}(x,t) = e^{-\beta t} \sin kx \,(A \cos \omega t + B \sin \omega t) \ .$$

Find k from the boundary conditions $u(0,t) = u(L,t) = 0$. Then use the PDE to find constraints on β, ω, A, and B. Set up a complete initial-boundary value problem and its solution.
Filename: `damped_waves`.

Problem 2.8: Explore symmetry boundary conditions
Consider the simple "plug" wave where $\Omega = [-L, L]$ and

$$I(x) = \begin{cases} 1, & x \in [-\delta, \delta], \\ 0, & \text{otherwise} \end{cases}$$

for some number $0 < \delta < L$. The other initial condition is $u_t(x,0) = 0$ and there is no source term f. The boundary conditions can be set to $u = 0$. The solution to this problem is symmetric around $x = 0$. This means that we can simulate the wave process in only half of the domain $[0, L]$.

a) Argue why the symmetry boundary condition is $u_x = 0$ at $x = 0$.

Hint Symmetry of a function about $x = x_0$ means that $f(x_0 + h) = f(x_0 - h)$.

b) Perform simulations of the complete wave problem on $[-L, L]$. Thereafter, utilize the symmetry of the solution and run a simulation in half of the domain $[0, L]$, using a boundary condition at $x = 0$. Compare plots from the two solutions and confirm that they are the same.
c) Prove the symmetry property of the solution by setting up the complete initial-boundary value problem and showing that if $u(x,t)$ is a solution, then also $u(-x,t)$ is a solution.
d) If the code works correctly, the solution $u(x,t) = x(L-x)(1+\frac{t}{2})$ should be reproduced exactly. Write a test function `test_quadratic` that checks whether this is the case. Simulate for x in $[0, \frac{L}{2}]$ with a symmetry condition at the end $x = \frac{L}{2}$.

Filename: `wave1D_symmetric`.

Exercise 2.9: Send pulse waves through a layered medium
Use the `pulse` function in `wave1D_dn_vc.py` to investigate sending a pulse, located with its peak at $x = 0$, through two media with different wave velocities. The

(scaled) velocity in the left medium is 1 while it is $\frac{1}{s_f}$ in the right medium. Report what happens with a Gaussian pulse, a "cosine hat" pulse, half a "cosine hat" pulse, and a plug pulse for resolutions $N_x = 40, 80, 160$, and $s_f = 2, 4$. Simulate until $T = 2$.
Filename: `pulse1D`.

Exercise 2.10: Explain why numerical noise occurs
The experiments performed in Exercise 2.9 shows considerable numerical noise in the form of non-physical waves, especially for $s_f = 4$ and the plug pulse or the half a "cosinehat" pulse. The noise is much less visible for a Gaussian pulse. Run the case with the plug and half a "cosinehat" pulse for $s_f = 1$, $C = 0.9, 0.25$, and $N_x = 40, 80, 160$. Use the numerical dispersion relation to explain the observations.
Filename: `pulse1D_analysis`.

Exercise 2.11: Investigate harmonic averaging in a 1D model
Harmonic means are often used if the wave velocity is non-smooth or discontinuous. Will harmonic averaging of the wave velocity give less numerical noise for the case $s_f = 4$ in Exercise 2.9?
Filename: `pulse1D_harmonic`.

Problem 2.12: Implement open boundary conditions
To enable a wave to leave the computational domain and travel undisturbed through the boundary $x = L$, one can in a one-dimensional problem impose the following condition, called a *radiation condition* or *open boundary condition*:

$$\frac{\partial u}{\partial t} + c\frac{\partial u}{\partial x} = 0\,. \tag{2.69}$$

The parameter c is the wave velocity.

Show that (2.69) accepts a solution $u = g_R(x - ct)$ (right-going wave), but not $u = g_L(x + ct)$ (left-going wave). This means that (2.69) will allow any right-going wave $g_R(x - ct)$ to pass through the boundary undisturbed.

A corresponding open boundary condition for a left-going wave through $x = 0$ is

$$\frac{\partial u}{\partial t} - c\frac{\partial u}{\partial x} = 0\,. \tag{2.70}$$

a) A natural idea for discretizing the condition (2.69) at the spatial end point $i = N_x$ is to apply centered differences in time and space:

$$[D_{2t}u + cD_{2x}u = 0]^n_i, \quad i = N_x\,. \tag{2.71}$$

Eliminate the fictitious value $u^n_{N_x+1}$ by using the discrete equation at the same point.
The equation for the first step, u^1_i, is in principle also affected, but we can then use the condition $u_{N_x} = 0$ since the wave has not yet reached the right boundary.

b) A much more convenient implementation of the open boundary condition at $x = L$ can be based on an explicit discretization

$$[D_t^+ u + cD_x^- u = 0]_i^n, \quad i = N_x \,. \tag{2.72}$$

From this equation, one can solve for $u_{N_x}^{n+1}$ and apply the formula as a Dirichlet condition at the boundary point. However, the finite difference approximations involved are of first order.
Implement this scheme for a wave equation $u_{tt} = c^2 u_{xx}$ in a domain $[0, L]$, where you have $u_x = 0$ at $x = 0$, the condition (2.69) at $x = L$, and an initial disturbance in the middle of the domain, e.g., a plug profile like

$$u(x,0) = \begin{cases} 1, & L/2 - \ell \le x \le L/2 + \ell, \\ 0, & \text{otherwise}\,. \end{cases}$$

Observe that the initial wave is split in two, the left-going wave is reflected at $x = 0$, and both waves travel out of $x = L$, leaving the solution as $u = 0$ in $[0, L]$. Use a unit Courant number such that the numerical solution is exact. Make a movie to illustrate what happens.
Because this simplified implementation of the open boundary condition works, there is no need to pursue the more complicated discretization in a).

Hint Modify the solver function in `wave1D_dn.py`.

c) Add the possibility to have either $u_x = 0$ or an open boundary condition at the left boundary. The latter condition is discretized as

$$[D_t^+ u - cD_x^+ u = 0]_i^n, \quad i = 0, \tag{2.73}$$

leading to an explicit update of the boundary value u_0^{n+1}.
The implementation can be tested with a Gaussian function as initial condition:

$$g(x; m, s) = \frac{1}{\sqrt{2\pi} s} e^{-\frac{(x-m)^2}{2s^2}} \,.$$

Run two tests:
(a) Disturbance in the middle of the domain, $I(x) = g(x; L/2, s)$, and open boundary condition at the left end.
(b) Disturbance at the left end, $I(x) = g(x; 0, s)$, and $u_x = 0$ as symmetry boundary condition at this end.
Make test functions for both cases, testing that the solution is zero after the waves have left the domain.

d) In 2D and 3D it is difficult to compute the correct wave velocity normal to the boundary, which is needed in generalizations of the open boundary conditions in higher dimensions. Test the effect of having a slightly wrong wave velocity in (2.72). Make movies to illustrate what happens.

Filename: `wave1D_open_BC`.

Remarks The condition (2.69) works perfectly in 1D when c is known. In 2D and 3D, however, the condition reads $u_t + c_x u_x + c_y u_y = 0$, where c_x and c_y are the wave speeds in the x and y directions. Estimating these components (i.e., the direction of the wave) is often challenging. Other methods are normally used in 2D and 3D to let waves move out of a computational domain.

Exercise 2.13: Implement periodic boundary conditions
It is frequently of interest to follow wave motion over large distances and long times. A straightforward approach is to work with a very large domain, but that might lead to a lot of computations in areas of the domain where the waves cannot be noticed. A more efficient approach is to let a right-going wave out of the domain and at the same time let it enter the domain on the left. This is called a *periodic boundary condition*.

The boundary condition at the right end $x = L$ is an open boundary condition (see Exercise 2.12) to let a right-going wave out of the domain. At the left end, $x = 0$, we apply, in the beginning of the simulation, either a symmetry boundary condition (see Exercise 2.8) $u_x = 0$, or an open boundary condition.

This initial wave will split in two and either be reflected or transported out of the domain at $x = 0$. The purpose of the exercise is to follow the right-going wave. We can do that with a *periodic boundary condition*. This means that when the right-going wave hits the boundary $x = L$, the open boundary condition lets the wave out of the domain, but at the same time we use a boundary condition on the left end $x = 0$ that feeds the outgoing wave into the domain again. This periodic condition is simply $u(0) = u(L)$. The switch from $u_x = 0$ or an open boundary condition at the left end to a periodic condition can happen when $u(L,t) > \epsilon$, where $\epsilon = 10^{-4}$ might be an appropriate value for determining when the right-going wave hits the boundary $x = L$.

The open boundary conditions can conveniently be discretized as explained in Exercise 2.12. Implement the described type of boundary conditions and test them on two different initial shapes: a plug $u(x,0) = 1$ for $x \leq 0.1$, $u(x,0) = 0$ for $x > 0.1$, and a Gaussian function in the middle of the domain: $u(x,0) = \exp\left(-\frac{1}{2}(x-0.5)^2/0.05\right)$. The domain is the unit interval $[0,1]$. Run these two shapes for Courant numbers 1 and 0.5. Assume constant wave velocity. Make movies of the four cases. Reason why the solutions are correct.
Filename: `periodic`.

Exercise 2.14: Compare discretizations of a Neumann condition
We have a 1D wave equation with variable wave velocity: $u_{tt} = (qu_x)_x$. A Neumann condition u_x at $x = 0, L$ can be discretized as shown in (2.54) and (2.57).

The aim of this exercise is to examine the rate of the numerical error when using different ways of discretizing the Neumann condition.

a) As a test problem, $q = 1 + (x - L/2)^4$ can be used, with $f(x,t)$ adapted such that the solution has a simple form, say $u(x,t) = \cos(\pi x/L)\cos(\omega t)$ for, e.g., $\omega = 1$. Perform numerical experiments and find the convergence rate of the error using the approximation (2.54).
b) Switch to $q(x) = 1 + \cos(\pi x/L)$, which is symmetric at $x = 0, L$, and check the convergence rate of the scheme (2.57). Now, $q_{i-1/2}$ is a 2nd-order approxi-

mation to q_i, $q_{i-1/2} = q_i + 0.25q_i''\Delta x^2 + \cdots$, because $q_i' = 0$ for $i = N_x$ (a similar argument can be applied to the case $i = 0$).

c) A third discretization can be based on a simple and convenient, but less accurate, one-sided difference: $u_i - u_{i-1} = 0$ at $i = N_x$ and $u_{i+1} - u_i = 0$ at $i = 0$. Derive the resulting scheme in detail and implement it. Run experiments with q from a) or b) to establish the rate of convergence of the scheme.

d) A fourth technique is to view the scheme as

$$[D_t D_t u]_i^n = \frac{1}{\Delta x}\left([qD_x u]_{i+\frac{1}{2}}^n - [qD_x u]_{i-\frac{1}{2}}^n\right) + [f]_i^n,$$

and place the boundary at $x_{i+\frac{1}{2}}$, $i = N_x$, instead of exactly at the physical boundary. With this idea of approximating (moving) the boundary, we can just set $[qD_x u]_{i+\frac{1}{2}}^n = 0$. Derive the complete scheme using this technique. The implementation of the boundary condition at $L - \Delta x/2$ is $\mathcal{O}(\Delta x^2)$ accurate, but the interesting question is what impact the movement of the boundary has on the convergence rate. Compute the errors as usual over the entire mesh and use q from a) or b).

Filename: `Neumann_discr`.

Exercise 2.15: Verification by a cubic polynomial in space

The purpose of this exercise is to verify the implementation of the `solver` function in the program `wave1D_n0.py` by using an exact numerical solution for the wave equation $u_{tt} = c^2 u_{xx} + f$ with Neumann boundary conditions $u_x(0,t) = u_x(L,t) = 0$.

A similar verification is used in the file `wave1D_u0.py`, which solves the same PDE, but with Dirichlet boundary conditions $u(0,t) = u(L,t) = 0$. The idea of the verification test in function `test_quadratic` in `wave1D_u0.py` is to produce a solution that is a lower-order polynomial such that both the PDE problem, the boundary conditions, and all the discrete equations are exactly fulfilled. Then the `solver` function should reproduce this exact solution to machine precision. More precisely, we seek $u = X(x)T(t)$, with $T(t)$ as a linear function and $X(x)$ as a parabola that fulfills the boundary conditions. Inserting this u in the PDE determines f. It turns out that u also fulfills the discrete equations, because the truncation error of the discretized PDE has derivatives in x and t of order four and higher. These derivatives all vanish for a quadratic $X(x)$ and linear $T(t)$.

It would be attractive to use a similar approach in the case of Neumann conditions. We set $u = X(x)T(t)$ and seek lower-order polynomials X and T. To force u_x to vanish at the boundary, we let X_x be a parabola. Then X is a cubic polynomial. The fourth-order derivative of a cubic polynomial vanishes, so $u = X(x)T(t)$ will fulfill the discretized PDE also in this case, if f is adjusted such that u fulfills the PDE.

However, the discrete boundary condition is not exactly fulfilled by this choice of u. The reason is that

$$[D_{2x}u]_i^n = u_x(x_i,t_n) + \frac{1}{6}u_{xxx}(x_i,t_n)\Delta x^2 + \mathcal{O}(\Delta x^4). \tag{2.74}$$

At the two boundary points, we must demand that the derivative $X_x(x) = 0$ such that $u_x = 0$. However, u_{xxx} is a constant and not zero when $X(x)$ is a cubic polynomial. Therefore, our $u = X(x)T(t)$ fulfills

$$[D_{2x}u]_i^n = \frac{1}{6}u_{xxx}(x_i, t_n)\Delta x^2,$$

and not

$$[D_{2x}u]_i^n = 0, \quad i = 0, N_x,$$

as it should. (Note that all the higher-order terms $\mathcal{O}(\Delta x^4)$ also have higher-order derivatives that vanish for a cubic polynomial.) So to summarize, the fundamental problem is that u as a product of a cubic polynomial and a linear or quadratic polynomial in time is not an exact solution of the discrete boundary conditions.

To make progress, we assume that $u = X(x)T(t)$, where T for simplicity is taken as a prescribed linear function $1 + \frac{1}{2}t$, and $X(x)$ is taken as an *unknown* cubic polynomial $\sum_{j=0}^{3} a_j x^j$. There are two different ways of determining the coefficients $a_0, \ldots, a_3$ such that both the discretized PDE and the discretized boundary conditions are fulfilled, under the constraint that we can specify a function $f(x,t)$ for the PDE to feed to the `solver` function in `wave1D_n0.py`. Both approaches are explained in the subexercises.

a) One can insert u in the discretized PDE and find the corresponding f. Then one can insert u in the discretized boundary conditions. This yields two equations for the four coefficients $a_0, \ldots, a_3$. To find the coefficients, one can set $a_0 = 0$ and $a_1 = 1$ for simplicity and then determine a_2 and a_3. This approach will make a_2 and a_3 depend on Δx and f will depend on both Δx and Δt.
Use `sympy` to perform analytical computations. A starting point is to define u as follows:

```
def test_cubic1():
    import sympy as sm
    x, t, c, L, dx, dt = sm.symbols('x t c L dx dt')
    i, n = sm.symbols('i n', integer=True)

    # Assume discrete solution is a polynomial of degree 3 in x
    T = lambda t: 1 + sm.Rational(1,2)*t  # Temporal term
    a = sm.symbols('a_0 a_1 a_2 a_3')
    X = lambda x: sum(a[q]*x**q for q in range(4))  # Spatial term
    u = lambda x, t: X(x)*T(t)
```

The symbolic expression for u is reached by calling `u(x,t)` with `x` and `t` as `sympy` symbols.
Define `DxDx(u, i, n)`, `DtDt(u, i, n)`, and `D2x(u, i, n)` as Python functions for returning the difference approximations $[D_xD_xu]_i^n$, $[D_tD_tu]_i^n$, and $[D_{2x}u]_i^n$. The next step is to set up the residuals for the equations $[D_{2x}u]_0^n = 0$ and $[D_{2x}u]_{N_x}^n = 0$, where $N_x = L/\Delta x$. Call the residuals `R_0` and `R_L`. Substitute a_0 and a_1 by 0 and 1, respectively, in `R_0`, `R_L`, and `a`:

```
R_0 = R_0.subs(a[0], 0).subs(a[1], 1)
R_L = R_L.subs(a[0], 0).subs(a[1], 1)
a = list(a)  # enable in-place assignment
a[0:2] = 0, 1
```

Determining a_2 and a_3 from the discretized boundary conditions is then about solving two equations with respect to a_2 and a_3, i.e., `a[2:]`:

```
s = sm.solve([R_0, R_L], a[2:])
# s is dictionary with the unknowns a[2] and a[3] as keys
a[2:] = s[a[2]], s[a[3]]
```

Now, a contains computed values and u will automatically use these new values since X accesses a.
Compute the source term f from the discretized PDE: $f_i^n = [D_tD_tu - c^2D_xD_xu]_i^n$. Turn u, the time derivative u_t (needed for the initial condition $V(x)$), and f into Python functions. Set numerical values for L, N_x, C, and c. Prescribe the time interval as $\Delta t = CL/(N_xc)$, which imply $\Delta x = c\Delta t/C = L/N_x$. Define new functions `I(x)`, `V(x)`, and `f(x,t)` as wrappers of the ones made above, where fixed values of L, c, Δx, and Δt are inserted, such that `I`, `V`, and `f` can be passed on to the `solver` function. Finally, call `solver` with a `user_action` function that compares the numerical solution to this exact solution u of the discrete PDE problem.

Hint To turn a `sympy` expression `e`, depending on a series of symbols, say `x`, `t`, `dx`, `dt`, `L`, and `c`, into a plain Python function `e_exact(x,t,L,dx,dt,c)`, one can write

```
e_exact = sm.lambdify([x,t,L,dx,dt,c], e, 'numpy')
```

The `'numpy'` argument is a good habit as the `e_exact` function will then work with array arguments if it contains mathematical functions (but here we only do plain arithmetics, which automatically work with arrays).

b) An alternative way of determining $a_0, \ldots, a_3$ is to reason as follows. We first construct $X(x)$ such that the boundary conditions are fulfilled: $X = x(L-x)$. However, to compensate for the fact that this choice of X does not fulfill the discrete boundary condition, we seek u such that

$$u_x = \frac{\partial}{\partial x}x(L-x)T(t) - \frac{1}{6}u_{xxx}\Delta x^2,$$

since this u will fit the discrete boundary condition. Assuming $u = T(t)\sum_{j=0}^{3} a_j x^j$, we can use the above equation to determine the coefficients a_1, a_2, a_3. A value, e.g., 1 can be used for a_0. The following `sympy` code computes this u:

```
def test_cubic2():
    import sympy as sm
    x, t, c, L, dx = sm.symbols('x t c L dx')
    T = lambda t: 1 + sm.Rational(1,2)*t  # Temporal term
    # Set u as a 3rd-degree polynomial in space
    X = lambda x: sum(a[i]*x**i for i in range(4))
    a = sm.symbols('a_0 a_1 a_2 a_3')
    u = lambda x, t: X(x)*T(t)
    # Force discrete boundary condition to be zero by adding
    # a correction term the analytical suggestion x*(L-x)*T
    # u_x = x*(L-x)*T(t) - 1/6*u_xxx*dx**2
    R = sm.diff(u(x,t), x) - (
        x*(L-x) - sm.Rational(1,6)*sm.diff(u(x,t), x, x, x)*dx**2)
    # R is a polynomial: force all coefficients to vanish.
    # Turn R to Poly to extract coefficients:
    R = sm.poly(R, x)
    coeff = R.all_coeffs()
    s = sm.solve(coeff, a[1:])  # a[0] is not present in R
    # s is dictionary with a[i] as keys
    # Fix a[0] as 1
    s[a[0]] = 1
    X = lambda x: sm.simplify(sum(s[a[i]]*x**i for i in range(4)))
    u = lambda x, t: X(x)*T(t)
    print 'u:', u(x,t)
```

The next step is to find the source term `f_e` by inserting `u_e` in the PDE. Thereafter, turn `u`, `f`, and the time derivative of `u` into plain Python functions as in a), and then wrap these functions in new functions `I`, `V`, and `f`, with the right signature as required by the `solver` function. Set parameters as in a) and check that the solution is exact to machine precision at each time level using an appropriate `user_action` function.

Filename: `wave1D_n0_test_cubic`.

2.10 Analysis of the Difference Equations

2.10.1 Properties of the Solution of the Wave Equation

The wave equation

$$\frac{\partial^2 u}{\partial t^2} = c^2 \frac{\partial^2 u}{\partial x^2}$$

has solutions of the form

$$u(x,t) = g_R(x - ct) + g_L(x + ct), \tag{2.75}$$

for any functions g_R and g_L sufficiently smooth to be differentiated twice. The result follows from inserting (2.75) in the wave equation. A function of the form $g_R(x - ct)$ represents a signal moving to the right in time with constant velocity c. This feature can be explained as follows. At time $t = 0$ the signal looks like $g_R(x)$. Introducing a moving horizontal coordinate $\xi = x - ct$, we see the function $g_R(\xi)$

is "at rest" in the ξ coordinate system, and the shape is always the same. Say the $g_R(\xi)$ function has a peak at $\xi = 0$. This peak is located at $x = ct$, which means that it moves with the velocity $dx/dt = c$ in the x coordinate system. Similarly, $g_L(x + ct)$ is a function, initially with shape $g_L(x)$, that moves in the negative x direction with constant velocity c (introduce $\xi = x + ct$, look at the point $\xi = 0$, $x = -ct$, which has velocity $dx/dt = -c$).

With the particular initial conditions

$$u(x,0) = I(x), \quad \frac{\partial}{\partial t}u(x,0) = 0,$$

we get, with u as in (2.75),

$$g_R(x) + g_L(x) = I(x), \quad -cg'_R(x) + cg'_L(x) = 0\,.$$

The former suggests $g_R = g_L$, and the former then leads to $g_R = g_L = I/2$. Consequently,

$$u(x,t) = \frac{1}{2}I(x - ct) + \frac{1}{2}I(x + ct)\,. \tag{2.76}$$

The interpretation of (2.76) is that the initial shape of u is split into two parts, each with the same shape as I but half of the initial amplitude. One part is traveling to the left and the other one to the right.

The solution has two important physical features: constant amplitude of the left and right wave, and constant velocity of these two waves. It turns out that the numerical solution will also preserve the constant amplitude, but the velocity depends on the mesh parameters Δt and Δx.

The solution (2.76) will be influenced by boundary conditions when the parts $\frac{1}{2}I(x - ct)$ and $\frac{1}{2}I(x + ct)$ hit the boundaries and get, e.g., reflected back into the domain. However, when $I(x)$ is nonzero only in a small part in the middle of the spatial domain $[0, L]$, which means that the boundaries are placed far away from the initial disturbance of u, the solution (2.76) is very clearly observed in a simulation.

A useful representation of solutions of wave equations is a linear combination of sine and/or cosine waves. Such a sum of waves is a solution if the governing PDE is linear and each sine or cosine wave fulfills the equation. To ease analytical calculations by hand we shall work with complex exponential functions instead of real-valued sine or cosine functions. The real part of complex expressions will typically be taken as the physical relevant quantity (whenever a physical relevant quantity is strictly needed). The idea now is to build $I(x)$ of complex wave components e^{ikx}:

$$I(x) \approx \sum_{k\in K} b_k e^{ikx}\,. \tag{2.77}$$

Here, k is the frequency of a component, K is some set of all the discrete k values needed to approximate $I(x)$ well, and b_k are constants that must be determined. We will very seldom need to compute the b_k coefficients: most of the insight we look for, and the understanding of the numerical methods we want to establish, come from investigating how the PDE and the scheme treat a single component e^{ikx} wave.

Letting the number of k values in K tend to infinity, makes the sum (2.77) converge to $I(x)$. This sum is known as a *Fourier series* representation of $I(x)$. Looking at (2.76), we see that the solution $u(x,t)$, when $I(x)$ is represented as in (2.77), is also built of basic complex exponential wave components of the form $e^{ik(x\pm ct)}$ according to

$$u(x,t) = \frac{1}{2}\sum_{k\in K} b_k e^{ik(x-ct)} + \frac{1}{2}\sum_{k\in K} b_k e^{ik(x+ct)} . \tag{2.78}$$

It is common to introduce the frequency in time $\omega = kc$ and assume that $u(x,t)$ is a sum of basic wave components written as $e^{ikx-\omega t}$. (Observe that inserting such a wave component in the governing PDE reveals that $\omega^2 = k^2c^2$, or $\omega = \pm kc$, reflecting the two solutions: one $(+kc)$ traveling to the right and the other $(-kc)$ traveling to the left.)

2.10.2 More Precise Definition of Fourier Representations

The above introduction to function representation by sine and cosine waves was quick and intuitive, but will suffice as background knowledge for the following material of single wave component analysis. However, to understand all details of how different wave components sum up to the analytical and numerical solutions, a more precise mathematical treatment is helpful and therefore summarized below.

It is well known that periodic functions can be represented by Fourier series. A generalization of the Fourier series idea to non-periodic functions defined on the real line is the *Fourier transform*:

$$I(x) = \int_{-\infty}^{\infty} A(k)e^{ikx}dk, \tag{2.79}$$

$$A(k) = \int_{-\infty}^{\infty} I(x)e^{-ikx}dx . \tag{2.80}$$

The function $A(k)$ reflects the weight of each wave component e^{ikx} in an infinite sum of such wave components. That is, $A(k)$ reflects the frequency content in the function $I(x)$. Fourier transforms are particularly fundamental for analyzing and understanding time-varying signals.

The solution of the linear 1D wave PDE can be expressed as

$$u(x,t) = \int_{-\infty}^{\infty} A(k)e^{i(kx-\omega(k)t)}dx .$$

In a finite difference method, we represent u by a mesh function u_q^n, where n counts temporal mesh points and q counts the spatial ones (the usual counter for spatial points, i, is here already used as imaginary unit). Similarly, $I(x)$ is approximated by the mesh function I_q, $q = 0,\ldots,N_x$. On a mesh, it does not make sense to work with wave components e^{ikx} for very large k, because the shortest possible sine or cosine wave that can be represented uniquely on a mesh with spacing Δx is the wave with wavelength $2\Delta x$. This wave has its peaks and throughs at every two mesh points. That is, the wave "jumps up and down" between the mesh points.

The corresponding k value for the shortest possible wave in the mesh is $k = 2\pi/(2\Delta x) = \pi/\Delta x$. This maximum frequency is known as the *Nyquist frequency*. Within the range of relevant frequencies $(0, \pi/\Delta x]$ one defines the discrete Fourier transform[11], using $N_x + 1$ discrete frequencies:

$$I_q = \frac{1}{N_x+1}\sum_{k=0}^{N_x} A_k e^{i2\pi kq/(N_x+1)}, \quad q = 0, \ldots, N_x, \tag{2.81}$$

$$A_k = \sum_{q=0}^{N_x} I_q e^{-i2\pi kq/(N_x+1)}, \quad k = 0, \ldots, N_x\,. \tag{2.82}$$

The A_k values represent the discrete Fourier transform of the I_q values, which themselves are the inverse discrete Fourier transform of the A_k values.

The discrete Fourier transform is efficiently computed by the *Fast Fourier transform* algorithm. For a real function $I(x)$, the relevant Python code for computing and plotting the discrete Fourier transform appears in the example below.

```
import numpy as np
from numpy import sin, pi

def I(x):
    return sin(2*pi*x) + 0.5*sin(4*pi*x) + 0.1*sin(6*pi*x)

# Mesh
L = 10; Nx = 100
x = np.linspace(0, L, Nx+1)
dx = L/float(Nx)

# Discrete Fourier transform
A = np.fft.rfft(I(x))
A_amplitude = np.abs(A)

# Compute the corresponding frequencies
freqs = np.linspace(0, pi/dx, A_amplitude.size)

import matplotlib.pyplot as plt
plt.plot(freqs, A_amplitude)
plt.show()
```

2.10.3 Stability

The scheme

$$[D_tD_tu = c^2D_xD_xu]^n_q \tag{2.83}$$

[11] http://en.wikipedia.org/wiki/Discrete_Fourier_transform

for the wave equation $u_{tt} = c^2 u_{xx}$ allows basic wave components

$$u_q^n = e^{i(kx_q - \tilde{\omega} t_n)}$$

as solution, but it turns out that the frequency in time, $\tilde{\omega}$, is not equal to the exact frequency $\omega = kc$. The goal now is to find exactly what $\tilde{\omega}$ is. We ask two key questions:

- How accurate is $\tilde{\omega}$ compared to ω?
- Does the amplitude of such a wave component preserve its (unit) amplitude, as it should, or does it get amplified or damped in time (because of a complex $\tilde{\omega}$)?

The following analysis will answer these questions. We shall continue using q as an identifier for a certain mesh point in the x direction.

Preliminary results A key result needed in the investigations is the finite difference approximation of a second-order derivative acting on a complex wave component:

$$[D_t D_t e^{i\omega t}]^n = -\frac{4}{\Delta t^2} \sin^2\left(\frac{\omega \Delta t}{2}\right) e^{i\omega n \Delta t}.$$

By just changing symbols ($\omega \to k$, $t \to x$, $n \to q$) it follows that

$$[D_x D_x e^{ikx}]_q = -\frac{4}{\Delta x^2} \sin^2\left(\frac{k \Delta x}{2}\right) e^{ikq\Delta x}.$$

Numerical wave propagation Inserting a basic wave component $u_q^n = e^{i(kx_q - \tilde{\omega} t_n)}$ in (2.83) results in the need to evaluate two expressions:

$$\begin{aligned} [D_t D_t e^{ikx} e^{-i\tilde{\omega} t}]_q^n &= [D_t D_t e^{-i\tilde{\omega} t}]^n e^{ikq\Delta x} \\ &= -\frac{4}{\Delta t^2} \sin^2\left(\frac{\tilde{\omega} \Delta t}{2}\right) e^{-i\tilde{\omega} n \Delta t} e^{ikq\Delta x} \end{aligned} \tag{2.84}$$

$$\begin{aligned} [D_x D_x e^{ikx} e^{-i\tilde{\omega} t}]_q^n &= [D_x D_x e^{ikx}]_q e^{-i\tilde{\omega} n \Delta t} \\ &= -\frac{4}{\Delta x^2} \sin^2\left(\frac{k \Delta x}{2}\right) e^{ikq\Delta x} e^{-i\tilde{\omega} n \Delta t}. \end{aligned} \tag{2.85}$$

Then the complete scheme,

$$[D_t D_t e^{ikx} e^{-i\tilde{\omega} t} = c^2 D_x D_x e^{ikx} e^{-i\tilde{\omega} t}]_q^n$$

leads to the following equation for the unknown numerical frequency $\tilde{\omega}$ (after dividing by $-e^{ikx} e^{-i\tilde{\omega} t}$):

$$\frac{4}{\Delta t^2} \sin^2\left(\frac{\tilde{\omega} \Delta t}{2}\right) = c^2 \frac{4}{\Delta x^2} \sin^2\left(\frac{k \Delta x}{2}\right),$$

or

$$\sin^2\left(\frac{\tilde{\omega} \Delta t}{2}\right) = C^2 \sin^2\left(\frac{k \Delta x}{2}\right), \tag{2.86}$$

where

$$C = \frac{c\Delta t}{\Delta x} \tag{2.87}$$

is the Courant number. Taking the square root of (2.86) yields

$$\sin\left(\frac{\tilde{\omega}\Delta t}{2}\right) = C \sin\left(\frac{k\Delta x}{2}\right). \tag{2.88}$$

Since the exact ω is real it is reasonable to look for a real solution $\tilde{\omega}$ of (2.88). The right-hand side of (2.88) must then be in $[-1, 1]$ because the sine function on the left-hand side has values in $[-1, 1]$ for real $\tilde{\omega}$. The magnitude of the sine function on the right-hand side attains the value 1 when

$$\frac{k\Delta x}{2} = \frac{\pi}{2} + m\pi, \quad m \in \mathbb{Z}.$$

With $m = 0$ we have $k\Delta x = \pi$, which means that the wavelength $\lambda = 2\pi/k$ becomes $2\Delta x$. This is the absolutely shortest wavelength that can be represented on the mesh: the wave jumps up and down between each mesh point. Larger values of $|m|$ are irrelevant since these correspond to k values whose waves are too short to be represented on a mesh with spacing Δx. For the shortest possible wave in the mesh, $\sin(k\Delta x/2) = 1$, and we must require

$$C \leq 1. \tag{2.89}$$

Consider a right-hand side in (2.88) of magnitude larger than unity. The solution $\tilde{\omega}$ of (2.88) must then be a complex number $\tilde{\omega} = \tilde{\omega}_r + i\tilde{\omega}_i$ because the sine function is larger than unity for a complex argument. One can show that for any ω_i there will also be a corresponding solution with $-\omega_i$. The component with $\omega_i > 0$ gives an amplification factor $e^{\omega_i t}$ that grows exponentially in time. We cannot allow this and must therefore require $C \leq 1$ as a *stability criterion*.

Remark on the stability requirement

For smoother wave components with longer wave lengths per length Δx, (2.89) can in theory be relaxed. However, small round-off errors are always present in a numerical solution and these vary arbitrarily from mesh point to mesh point and can be viewed as unavoidable noise with wavelength $2\Delta x$. As explained, $C > 1$ will for this very small noise lead to exponential growth of the shortest possible wave component in the mesh. This noise will therefore grow with time and destroy the whole solution.

2.10.4 Numerical Dispersion Relation

Equation (2.88) can be solved with respect to $\tilde{\omega}$:

$$\tilde{\omega} = \frac{2}{\Delta t}\sin^{-1}\left(C \sin\left(\frac{k\Delta x}{2}\right)\right). \tag{2.90}$$

The relation between the numerical frequency $\tilde{\omega}$ and the other parameters k, c, Δx, and Δt is called a *numerical dispersion relation*. Correspondingly, $\omega = kc$ is the *analytical dispersion relation*. In general, dispersion refers to the phenomenon where the wave velocity depends on the spatial frequency (k, or the wave length $\lambda = 2\pi/k$) of the wave. Since the wave velocity is $\omega/k = c$, we realize that the analytical dispersion relation reflects the fact that there is no dispersion. However, in a numerical scheme we have dispersive waves where the wave velocity depends on k.

The special case $C = 1$ deserves attention since then the right-hand side of (2.90) reduces to

$$\frac{2}{\Delta t}\frac{k\Delta x}{2} = \frac{1}{\Delta t}\frac{\omega\Delta x}{c} = \frac{\omega}{C} = \omega\,.$$

That is, $\tilde{\omega} = \omega$ and the numerical solution is exact at all mesh points regardless of Δx and Δt! This implies that the numerical solution method is also an analytical solution method, at least for computing u at discrete points (the numerical method says nothing about the variation of u *between* the mesh points, and employing the common linear interpolation for extending the discrete solution gives a curve that in general deviates from the exact one).

For a closer examination of the error in the numerical dispersion relation when $C < 1$, we can study $\tilde{\omega} - \omega$, $\tilde{\omega}/\omega$, or the similar error measures in wave velocity: $\tilde{c} - c$ and $\tilde{c}/c$, where $c = \omega/k$ and $\tilde{c} = \tilde{\omega}/k$. It appears that the most convenient expression to work with is $\tilde{c}/c$, since it can be written as a function of just two parameters:

$$\frac{\tilde{c}}{c} = \frac{1}{Cp}\sin^{-1}(C\sin p)\,,$$

with $p = k\Delta x/2$ as a non-dimensional measure of the spatial frequency. In essence, p tells how many spatial mesh points we have per wave length in space for the wave component with frequency k (recall that the wave length is $2\pi/k$). That is, p reflects how well the spatial variation of the wave component is resolved in the mesh. Wave components with wave length less than $2\Delta x$ ($2\pi/k < 2\Delta x$) are not visible in the mesh, so it does not make sense to have $p > \pi/2$.

We may introduce the function $r(C, p) = \tilde{c}/c$ for further investigation of numerical errors in the wave velocity:

$$r(C, p) = \frac{1}{Cp}\sin^{-1}(C\sin p)\,, \quad C \in (0, 1],\ p \in (0, \pi/2]\,. \tag{2.91}$$

This function is very well suited for plotting since it combines several parameters in the problem into a dependence on two dimensionless numbers, C and p.

Defining

```
def r(C, p):
    return 2/(C*p)*asin(C*sin(p))
```

we can plot $r(C, p)$ as a function of p for various values of C, see Fig. 2.6. Note that the shortest waves have the most erroneous velocity, and that short waves move more slowly than they should.

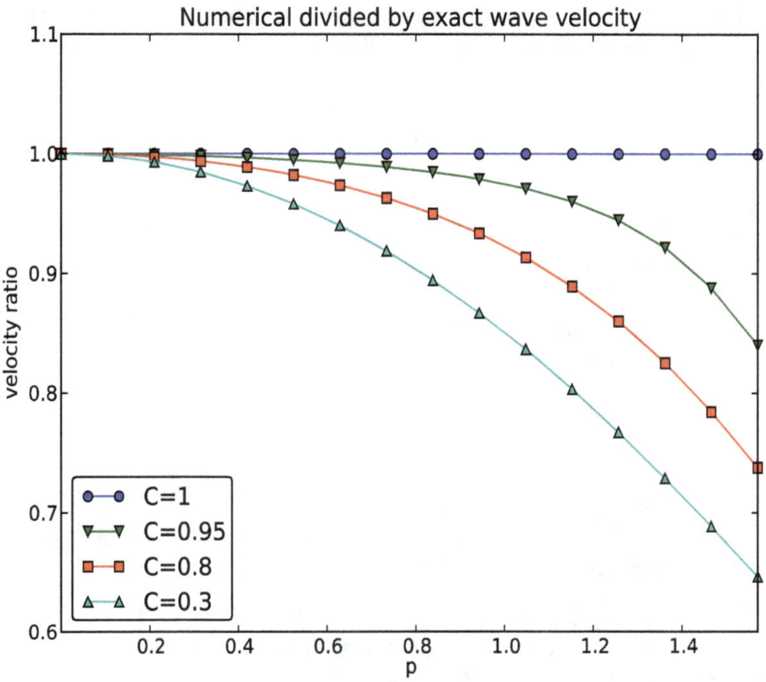

Fig. 2.6 The fractional error in the wave velocity for different Courant numbers

We can also easily make a Taylor series expansion in the discretization parameter p:

```
>>> import sympy as sym
>>> C, p = sym.symbols('C p')
>>> # Compute the 7 first terms around p=0 with no O() term
>>> rs = r(C, p).series(p, 0, 7).removeO()
>>> rs
p**6*(5*C**6/112 - C**4/16 + 13*C**2/720 - 1/5040) +
p**4*(3*C**4/40 - C**2/12 + 1/120) +
p**2*(C**2/6 - 1/6) + 1

>>> # Pick out the leading order term, but drop the constant 1
>>> rs_error_leading_order = (rs - 1).extract_leading_order(p)
>>> rs_error_leading_order
p**2*(C**2/6 - 1/6)

>>> # Turn the series expansion into a Python function
>>> rs_pyfunc = lambdify([C, p], rs, modules='numpy')

>>> # Check: rs_pyfunc is exact (=1) for C=1
>>> rs_pyfunc(1, 0.1)
1.0
```

Note that without the `.removeO()` call the series gets an `O(x**7)` term that makes it impossible to convert the series to a Python function (for, e.g., plotting).

From the `rs_error_leading_order` expression above, we see that the leading order term in the error of this series expansion is

$$\frac{1}{6}\left(\frac{k\Delta x}{2}\right)^2 (C^2-1) = \frac{k^2}{24}\left(c^2\Delta t^2 - \Delta x^2\right), \tag{2.92}$$

pointing to an error $\mathcal{O}(\Delta t^2, \Delta x^2)$, which is compatible with the errors in the difference approximations (D_tD_tu and D_xD_xu).

We can do more with a series expansion, e.g., factor it to see how the factor $C-1$ plays a significant role. To this end, we make a list of the terms, factor each term, and then sum the terms:

```
>>> rs = r(C, p).series(p, 0, 4).removeO().as_ordered_terms()
>>> rs
[1, C**2*p**2/6 - p**2/6,
 3*C**4*p**4/40 - C**2*p**4/12 + p**4/120,
 5*C**6*p**6/112 - C**4*p**6/16 + 13*C**2*p**6/720 - p**6/5040]
>>> rs = [factor(t) for t in rs]
>>> rs
[1, p**2*(C - 1)*(C + 1)/6,
 p**4*(C - 1)*(C + 1)*(3*C - 1)*(3*C + 1)/120,
 p**6*(C - 1)*(C + 1)*(225*C**4 - 90*C**2 + 1)/5040]
>>> rs = sum(rs)  # Python's sum function sums the list
>>> rs
p**6*(C - 1)*(C + 1)*(225*C**4 - 90*C**2 + 1)/5040 +
p**4*(C - 1)*(C + 1)*(3*C - 1)*(3*C + 1)/120 +
p**2*(C - 1)*(C + 1)/6 + 1
```

We see from the last expression that $C = 1$ makes all the terms in `rs` vanish. Since we already know that the numerical solution is exact for $C = 1$, the remaining terms in the Taylor series expansion will also contain factors of $C - 1$ and cancel for $C = 1$.

2.10.5 Extending the Analysis to 2D and 3D

The typical analytical solution of a 2D wave equation

$$u_{tt} = c^2(u_{xx} + u_{yy}),$$

is a wave traveling in the direction of $\boldsymbol{k} = k_x\boldsymbol{i} + k_y\boldsymbol{j}$, where $\boldsymbol{i}$ and $\boldsymbol{j}$ are unit vectors in the x and y directions, respectively ($\boldsymbol{i}$ should not be confused with $i = \sqrt{-1}$ here). Such a wave can be expressed by

$$u(x,y,t) = g(k_xx + k_yy - kct)$$

for some twice differentiable function g, or with $\omega = kc$, $k = |\boldsymbol{k}|$:

$$u(x,y,t) = g(k_xx + k_yy - \omega t)\,.$$

We can, in particular, build a solution by adding complex Fourier components of the form

$$e^{\left(i(k_x x+k_y y-\omega t)\right)}.$$

A discrete 2D wave equation can be written as

$$[D_t D_t u = c^2(D_x D_x u + D_y D_y u)]^n_{q,r}. \tag{2.93}$$

This equation admits a Fourier component

$$u^n_{q,r} = e^{\left(i(k_x q\Delta x+k_y r\Delta y-\tilde{\omega} n\Delta t)\right)}, \tag{2.94}$$

as solution. Letting the operators $D_t D_t$, $D_x D_x$, and $D_y D_y$ act on $u^n_{q,r}$ from (2.94) transforms (2.93) to

$$\frac{4}{\Delta t^2}\sin^2\left(\frac{\tilde{\omega}\Delta t}{2}\right) = c^2\frac{4}{\Delta x^2}\sin^2\left(\frac{k_x\Delta x}{2}\right) + c^2\frac{4}{\Delta y^2}\sin^2\left(\frac{k_y\Delta y}{2}\right) \tag{2.95}$$

or

$$\sin^2\left(\frac{\tilde{\omega}\Delta t}{2}\right) = C_x^2\sin^2 p_x + C_y^2\sin^2 p_y, \tag{2.96}$$

where we have eliminated the factor 4 and introduced the symbols

$$C_x = \frac{c\Delta t}{\Delta x}, \quad C_y = \frac{c\Delta t}{\Delta y}, \quad p_x = \frac{k_x\Delta x}{2}, \quad p_y = \frac{k_y\Delta y}{2}.$$

For a real-valued $\tilde{\omega}$ the right-hand side must be less than or equal to unity in absolute value, requiring in general that

$$C_x^2 + C_y^2 \le 1. \tag{2.97}$$

This gives the stability criterion, more commonly expressed directly in an inequality for the time step:

$$\Delta t \le \frac{1}{c}\left(\frac{1}{\Delta x^2}+\frac{1}{\Delta y^2}\right)^{-1/2}. \tag{2.98}$$

A similar, straightforward analysis for the 3D case leads to

$$\Delta t \le \frac{1}{c}\left(\frac{1}{\Delta x^2}+\frac{1}{\Delta y^2}+\frac{1}{\Delta z^2}\right)^{-1/2}. \tag{2.99}$$

In the case of a variable coefficient $c^2 = c^2(\boldsymbol{x})$, we must use the worst-case value

$$\bar{c} = \sqrt{\max_{\boldsymbol{x}\in\Omega} c^2(\boldsymbol{x})} \tag{2.100}$$

in the stability criteria. Often, especially in the variable wave velocity case, it is wise to introduce a safety factor $\beta \in (0, 1]$ too:

$$\Delta t \le \beta\frac{1}{\bar{c}}\left(\frac{1}{\Delta x^2}+\frac{1}{\Delta y^2}+\frac{1}{\Delta z^2}\right)^{-1/2}. \tag{2.101}$$

The exact numerical dispersion relations in 2D and 3D becomes, for constant c,

$$\tilde{\omega} = \frac{2}{\Delta t}\sin^{-1}\left(\left(C_x^2\sin^2 p_x + C_y^2\sin^2 p_y\right)^{\frac{1}{2}}\right), \tag{2.102}$$

$$\tilde{\omega} = \frac{2}{\Delta t}\sin^{-1}\left(\left(C_x^2\sin^2 p_x + C_y^2\sin^2 p_y + C_z^2\sin^2 p_z\right)^{\frac{1}{2}}\right). \tag{2.103}$$

We can visualize the numerical dispersion error in 2D much like we did in 1D. To this end, we need to reduce the number of parameters in $\tilde{\omega}$. The direction of the wave is parameterized by the polar angle θ, which means that

$$k_x = k\sin\theta,\quad k_y = k\cos\theta\,.$$

A simplification is to set $\Delta x = \Delta y = h$. Then $C_x = C_y = c\Delta t/h$, which we call C. Also,

$$p_x = \frac{1}{2}kh\cos\theta,\quad p_y = \frac{1}{2}kh\sin\theta\,.$$

The numerical frequency $\tilde{\omega}$ is now a function of three parameters:

- C, reflecting the number of cells a wave is displaced during a time step,
- $p = \frac{1}{2}kh$, reflecting the number of cells per wave length in space,
- θ, expressing the direction of the wave.

We want to visualize the error in the numerical frequency. To avoid having Δt as a free parameter in $\tilde{\omega}$, we work with $\tilde{c}/c = \tilde{\omega}/(kc)$. The coefficient in front of the $\sin^{-1}$ factor is then

$$\frac{2}{kc\Delta t} = \frac{2}{2kc\Delta th/h} = \frac{1}{Ckh} = \frac{2}{Cp},$$

and

$$\frac{\tilde{c}}{c} = \frac{2}{Cp}\sin^{-1}\left(C\left(\sin^2(p\cos\theta) + \sin^2(p\sin\theta)\right)^{\frac{1}{2}}\right).$$

We want to visualize this quantity as a function of p and θ for some values of $C \leq 1$. It is instructive to make color contour plots of $1 - \tilde{c}/c$ in *polar coordinates* with θ as the angular coordinate and p as the radial coordinate.

The stability criterion (2.97) becomes $C \leq C_{\max} = 1/\sqrt{2}$ in the present 2D case with the C defined above. Let us plot $1 - \tilde{c}/c$ in polar coordinates for $C_{\max}, 0.9C_{\max}, 0.5C_{\max}, 0.2C_{\max}$. The program below does the somewhat tricky work in Matplotlib, and the result appears in Fig. 2.7. From the figure we clearly see that the maximum C value gives the best results, and that waves whose propagation direction makes an angle of 45 degrees with an axis are the most accurate.

```
def dispersion_relation_2D(p, theta, C):
    arg = C*sqrt(sin(p*cos(theta))**2 +
                 sin(p*sin(theta))**2)
    c_frac = 2./(C*p)*arcsin(arg)

    return c_frac

import numpy as np
from numpy import \
     cos, sin, arcsin, sqrt, pi  # for nicer math formulas

r = p = np.linspace(0.001, pi/2, 101)
theta = np.linspace(0, 2*pi, 51)
r, theta = np.meshgrid(r, theta)

# Make 2x2 filled contour plots for 4 values of C
import matplotlib.pyplot as plt
C_max = 1/sqrt(2)
C = [[C_max, 0.9*C_max], [0.5*C_max, 0.2*C_max]]
fix, axes = plt.subplots(2, 2, subplot_kw=dict(polar=True))
for row in range(2):
    for column in range(2):
        error = 1 - dispersion_relation_2D(
            p, theta, C[row][column])
        print error.min(), error.max()
        # use vmin=error.min(), vmax=error.max()
        cax = axes[row][column].contourf(
            theta, r, error, 50, vmin=-1, vmax=-0.28)
        axes[row][column].set_xticks([])
        axes[row][column].set_yticks([])

# Add colorbar to the last plot
cbar = plt.colorbar(cax)
cbar.ax.set_ylabel('error in wave velocity')
plt.savefig('disprel2D.png');  plt.savefig('disprel2D.pdf')
plt.show()
```

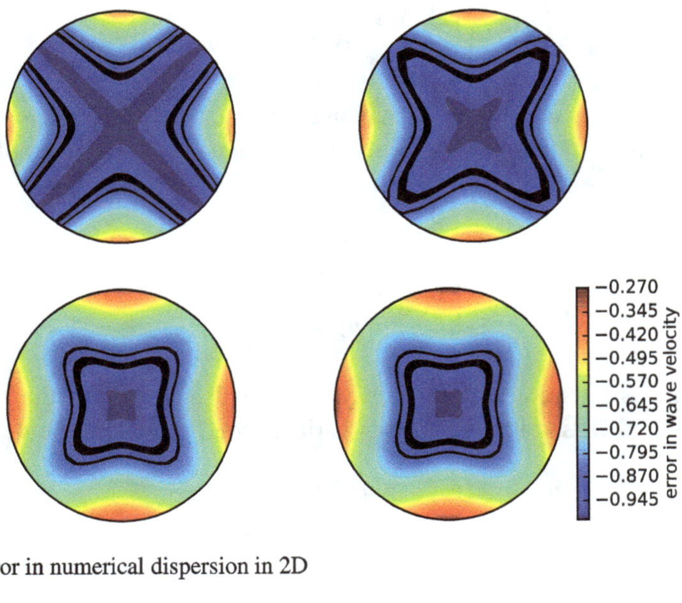

Fig. 2.7 Error in numerical dispersion in 2D

2.11 Finite Difference Methods for 2D and 3D Wave Equations

A natural next step is to consider extensions of the methods for various variants of the one-dimensional wave equation to two-dimensional (2D) and three-dimensional (3D) versions of the wave equation.

2.11.1 Multi-Dimensional Wave Equations

The general wave equation in d space dimensions, with constant wave velocity c, can be written in the compact form

$$\frac{\partial^2 u}{\partial t^2} = c^2\nabla^2 u \text{ for } \boldsymbol{x} \in \Omega \subset \mathbb{R}^d, \quad t \in (0, T], \tag{2.104}$$

where

$$\nabla^2 u = \frac{\partial^2 u}{\partial x^2} + \frac{\partial^2 u}{\partial y^2},$$

in a 2D problem ($d = 2$) and

$$\nabla^2 u = \frac{\partial^2 u}{\partial x^2} + \frac{\partial^2 u}{\partial y^2} + \frac{\partial^2 u}{\partial z^2},$$

in three space dimensions ($d = 3$).

Many applications involve variable coefficients, and the general wave equation in d dimensions is in this case written as

$$\varrho\frac{\partial^2 u}{\partial t^2} = \nabla\cdot(q\nabla u) + f \text{ for } \boldsymbol{x} \in \Omega \subset \mathbb{R}^d, \quad t \in (0, T], \tag{2.105}$$

which in, e.g., 2D becomes

$$\varrho(x,y)\frac{\partial^2 u}{\partial t^2} = \frac{\partial}{\partial x}\left(q(x,y)\frac{\partial u}{\partial x}\right) + \frac{\partial}{\partial y}\left(q(x,y)\frac{\partial u}{\partial y}\right) + f(x,y,t)\,. \tag{2.106}$$

To save some writing and space we may use the index notation, where subscript t, x, or y means differentiation with respect to that coordinate. For example,

$$\begin{aligned}\frac{\partial^2 u}{\partial t^2} &= u_{tt},\\ \frac{\partial}{\partial y}\left(q(x,y)\frac{\partial u}{\partial y}\right) &= (qu_y)_y\,.\end{aligned}$$

These comments extend straightforwardly to 3D, which means that the 3D versions of the two wave PDEs, with and without variable coefficients, can be stated as

$$u_{tt} = c^2(u_{xx} + u_{yy} + u_{zz}) + f, \tag{2.107}$$

$$\varrho u_{tt} = (qu_x)_x + (qu_y)_y + (qu_z)_z + f\,. \tag{2.108}$$

At *each point* of the boundary $\partial\Omega$ (of Ω) we need *one* boundary condition involving the unknown u. The boundary conditions are of three principal types:

1. u is prescribed ($u = 0$ or a known time variation of u at the boundary points, e.g., modeling an incoming wave),
2. $\partial u/\partial n = \boldsymbol{n} \cdot \nabla u$ is prescribed (zero for reflecting boundaries),
3. an open boundary condition (also called radiation condition) is specified to let waves travel undisturbed out of the domain, see Exercise 2.12 for details.

All the listed wave equations with *second-order* derivatives in time need *two* initial conditions:

1. $u = I$,
2. $u_t = V$.

2.11.2 Mesh

We introduce a mesh in time and in space. The mesh in time consists of time points

$$t_0 = 0 < t_1 < \cdots < t_{N_t},$$

normally, for wave equation problems, with a constant spacing $\Delta t = t_{n+1} - t_n$, $n \in \mathcal{I}_t^-$.

Finite difference methods are easy to implement on simple rectangle- or box-shaped *spatial domains*. More complicated shapes of the spatial domain require substantially more advanced techniques and implementational efforts (and a finite element method is usually a more convenient approach). On a rectangle- or box-shaped domain, mesh points are introduced separately in the various space directions:

$$\begin{aligned} x_0 < x_1 < \cdots < x_{N_x} & \text{ in the } x \text{ direction}, \\ y_0 < y_1 < \cdots < y_{N_y} & \text{ in the } y \text{ direction}, \\ z_0 < z_1 < \cdots < z_{N_z} & \text{ in the } z \text{ direction}. \end{aligned}$$

We can write a general mesh point as (x_i, y_j, z_k, t_n), with $i \in \mathcal{I}_x$, $j \in \mathcal{I}_y$, $k \in \mathcal{I}_z$, and $n \in \mathcal{I}_t$.

It is a very common choice to use constant mesh spacings: $\Delta x = x_{i+1} - x_i$, $i \in \mathcal{I}_x^-$, $\Delta y = y_{j+1} - y_j$, $j \in \mathcal{I}_y^-$, and $\Delta z = z_{k+1} - z_k$, $k \in \mathcal{I}_z^-$. With equal mesh spacings one often introduces $h = \Delta x = \Delta y = \Delta z$.

The unknown u at mesh point (x_i, y_j, z_k, t_n) is denoted by $u^n_{i,j,k}$. In 2D problems we just skip the z coordinate (by assuming no variation in that direction: $\partial/\partial z = 0$) and write $u^n_{i,j}$.

2.11.3 Discretization

Two- and three-dimensional wave equations are easily discretized by assembling building blocks for discretization of 1D wave equations, because the multi-dimensional versions just contain terms of the same type as those in 1D.

Discretizing the PDEs Equation (2.107) can be discretized as

$$[D_tD_tu = c^2(D_xD_xu + D_yD_yu + D_zD_zu) + f]^n_{i,j,k}\,. \tag{2.109}$$

A 2D version might be instructive to write out in detail:

$$[D_tD_tu = c^2(D_xD_xu + D_yD_yu) + f]^n_{i,j},$$

which becomes

$$\frac{u^{n+1}_{i,j} - 2u^n_{i,j} + u^{n-1}_{i,j}}{\Delta t^2} = c^2\frac{u^n_{i+1,j} - 2u^n_{i,j} + u^n_{i-1,j}}{\Delta x^2} + c^2\frac{u^n_{i,j+1} - 2u^n_{i,j} + u^n_{i,j-1}}{\Delta y^2} + f^n_{i,j}\,.$$

Assuming, as usual, that all values at time levels n and $n-1$ are known, we can solve for the only unknown $u^{n+1}_{i,j}$. The result can be compactly written as

$$u^{n+1}_{i,j} = 2u^n_{i,j} + u^{n-1}_{i,j} + c^2\Delta t^2[D_xD_xu + D_yD_yu]^n_{i,j}\,. \tag{2.110}$$

As in the 1D case, we need to develop a special formula for $u^1_{i,j}$ where we combine the general scheme for $u^{n+1}_{i,j}$, when $n = 0$, with the discretization of the initial condition:

$$[D_{2t}u = V]^0_{i,j} \quad\Rightarrow\quad u^{-1}_{i,j} = u^1_{i,j} - 2\Delta t V_{i,j}\,.$$

The result becomes, in compact form,

$$u^1_{i,j} = u^0_{i,j} - 2\Delta V_{i,j} + \frac{1}{2}c^2\Delta t^2[D_xD_xu + D_yD_yu]^0_{i,j}\,. \tag{2.111}$$

The PDE (2.108) with variable coefficients is discretized term by term using the corresponding elements from the 1D case:

$$[\varrho D_tD_tu = (D_x\overline{q}^xD_xu + D_y\overline{q}^yD_yu + D_z\overline{q}^zD_zu) + f]^n_{i,j,k}\,. \tag{2.112}$$

When written out and solved for the unknown $u^{n+1}_{i,j,k}$, one gets the scheme

$$
\begin{aligned}
u^{n+1}_{i,j,k} = & -u^{n-1}_{i,j,k} + 2u^{n}_{i,j,k} \\
& + \frac{1}{\varrho_{i,j,k}}\frac{1}{\Delta x^2}\left(\frac{1}{2}(q_{i,j,k}+q_{i+1,j,k})(u^n_{i+1,j,k}-u^n_{i,j,k})\right. \\
& \left. -\frac{1}{2}(q_{i-1,j,k}+q_{i,j,k})(u^n_{i,j,k}-u^n_{i-1,j,k})\right) \\
& + \frac{1}{\varrho_{i,j,k}}\frac{1}{\Delta y^2}\left(\frac{1}{2}(q_{i,j,k}+q_{i,j+1,k})(u^n_{i,j+1,k}-u^n_{i,j,k})\right. \\
& \left. -\frac{1}{2}(q_{i,j-1,k}+q_{i,j,k})(u^n_{i,j,k}-u^n_{i,j-1,k})\right) \\
& + \frac{1}{\varrho_{i,j,k}}\frac{1}{\Delta z^2}\left(\frac{1}{2}(q_{i,j,k}+q_{i,j,k+1})(u^n_{i,j,k+1}-u^n_{i,j,k})\right. \\
& \left. -\frac{1}{2}(q_{i,j,k-1}+q_{i,j,k})(u^n_{i,j,k}-u^n_{i,j,k-1})\right) \\
& + \Delta t^2 f^n_{i,j,k}\,.
\end{aligned}
$$

Also here we need to develop a special formula for $u^1_{i,j,k}$ by combining the scheme for $n = 0$ with the discrete initial condition, which is just a matter of inserting $u^{-1}_{i,j,k} = u^1_{i,j,k} - 2\Delta t V_{i,j,k}$ in the scheme and solving for $u^1_{i,j,k}$.

Handling boundary conditions where *u* is known The schemes listed above are valid for the internal points in the mesh. After updating these, we need to visit all the mesh points at the boundaries and set the prescribed u value.

Discretizing the Neumann condition The condition $\partial u/\partial n = 0$ was implemented in 1D by discretizing it with a $D_{2x}u$ centered difference, followed by eliminating the fictitious u point outside the mesh by using the general scheme at the boundary point. Alternatively, one can introduce ghost cells and update a ghost value for use in the Neumann condition. Exactly the same ideas are reused in multiple dimensions.

Consider the condition $\partial u/\partial n = 0$ at a boundary $y = 0$ of a rectangular domain $[0, L_x] \times [0, L_y]$ in 2D. The normal direction is then in $-y$ direction, so

$$
\frac{\partial u}{\partial n} = -\frac{\partial u}{\partial y},
$$

and we set

$$
[-D_{2y}u = 0]^n_{i,0} \quad \Rightarrow \quad \frac{u^n_{i,1}-u^n_{i,-1}}{2\Delta y} = 0\,.
$$

From this it follows that $u^n_{i,-1} = u^n_{i,1}$. The discretized PDE at the boundary point $(i, 0)$ reads

$$
\frac{u^{n+1}_{i,0}-2u^n_{i,0}+u^{n-1}_{i,0}}{\Delta t^2} = c^2\frac{u^n_{i+1,0}-2u^n_{i,0}+u^n_{i-1,0}}{\Delta x^2} + c^2\frac{u^n_{i,1}-2u^n_{i,0}+u^n_{i,-1}}{\Delta y^2} + f^n_{i,j}\,.
$$

We can then just insert $u^n_{i,1}$ for $u^n_{i,-1}$ in this equation and solve for the boundary value $u^{n+1}_{i,0}$, just as was done in 1D.

From these calculations, we see a pattern: the general scheme applies at the boundary $j = 0$ too if we just replace $j - 1$ by $j + 1$. Such a pattern is particularly useful for implementations. The details follow from the explained 1D case in Sect. 2.6.3.

The alternative approach to eliminating fictitious values outside the mesh is to have $u^n_{i,-1}$ available as a ghost value. The mesh is extended with one extra line (2D) or plane (3D) of ghost cells at a Neumann boundary. In the present example it means that we need a line with ghost cells below the y axis. The ghost values must be updated according to $u^{n+1}_{i,-1} = u^{n+1}_{i,1}$.

2.12 Implementation

We shall now describe in detail various Python implementations for solving a standard 2D, linear wave equation with constant wave velocity and $u = 0$ on the boundary. The wave equation is to be solved in the space-time domain $\Omega \times (0, T]$, where $\Omega = (0, L_x) \times (0, L_y)$ is a rectangular spatial domain. More precisely, the complete initial-boundary value problem is defined by

$$u_{tt} = c^2(u_{xx} + u_{yy}) + f(x, y, t), \quad (x, y) \in \Omega,\ t \in (0, T], \tag{2.113}$$

$$u(x, y, 0) = I(x, y), \quad (x, y) \in \Omega, \tag{2.114}$$

$$u_t(x, y, 0) = V(x, y), \quad (x, y) \in \Omega, \tag{2.115}$$

$$u = 0, \quad (x, y) \in \partial\Omega,\ t \in (0, T], \tag{2.116}$$

where $\partial\Omega$ is the boundary of Ω, in this case the four sides of the rectangle $\Omega = [0, L_x] \times [0, L_y]$: $x = 0$, $x = L_x$, $y = 0$, and $y = L_y$.

The PDE is discretized as

$$[D_t D_t u = c^2(D_x D_x u + D_y D_y u) + f]^n_{i,j},$$

which leads to an explicit updating formula to be implemented in a program:

$$\begin{aligned} u^{n+1}_{i,j} = &-u^{n-1}_{i,j} + 2u^n_{i,j} \\ &+ C_x^2(u^n_{i+1,j} - 2u^n_{i,j} + u^n_{i-1,j}) + C_y^2(u^n_{i,j+1} - 2u^n_{i,j} + u^n_{i,j-1}) \\ &+ \Delta t^2 f^n_{i,j}, \end{aligned} \tag{2.117}$$

for all interior mesh points $i \in \mathcal{I}^i_x$ and $j \in \mathcal{I}^i_y$, for $n \in \mathcal{I}^+_t$. The constants C_x and C_y are defined as

$$C_x = c\frac{\Delta t}{\Delta x}, \quad C_y = c\frac{\Delta t}{\Delta y}.$$

At the boundary, we simply set $u^{n+1}_{i,j} = 0$ for $i = 0$, $j = 0, \ldots, N_y$; $i = N_x$, $j = 0, \ldots, N_y$; $j = 0$, $i = 0, \ldots, N_x$; and $j = N_y$, $i = 0, \ldots, N_x$. For the first step, $n = 0$, (2.117) is combined with the discretization of the initial condition

$u_t = V$, $[D_{2t}u = V]^0_{i,j}$ to obtain a special formula for $u^1_{i,j}$ at the interior mesh points:

$$
\begin{aligned}
u^1_{i,j} &= u^0_{i,j} + \Delta t V_{i,j} \\
&\quad + \frac{1}{2}C_x^2(u^0_{i+1,j} - 2u^0_{i,j} + u^0_{i-1,j}) + \frac{1}{2}C_y^2(u^0_{i,j+1} - 2u^0_{i,j} + u^0_{i,j-1}) \\
&\quad + \frac{1}{2}\Delta t^2 f^n_{i,j} \,.
\end{aligned}
\tag{2.118}
$$

The algorithm is very similar to the one in 1D:

1. Set initial condition $u^0_{i,j} = I(x_i, y_j)$
2. Compute $u^1_{i,j}$ from (2.117)
3. Set $u^1_{i,j} = 0$ for the boundaries $i = 0, N_x$, $j = 0, N_y$
4. For $n = 1, 2, \ldots, N_t$:
   (a) Find $u^{n+1}_{i,j}$ from (2.117) for all internal mesh points, $i \in \mathcal{I}^i_x$, $j \in \mathcal{I}^i_y$
   (b) Set $u^{n+1}_{i,j} = 0$ for the boundaries $i = 0, N_x$, $j = 0, N_y$

2.12.1 Scalar Computations

The `solver` function for a 2D case with constant wave velocity and boundary condition $u = 0$ is analogous to the 1D case with similar parameter values (see `wave1D_u0.py`), apart from a few necessary extensions. The code is found in the program `wave2D_u0.py`.

Domain and mesh The spatial domain is now $[0, L_x] \times [0, L_y]$, specified by the arguments `Lx` and `Ly`. Similarly, the number of mesh points in the x and y directions, N_x and N_y, become the arguments `Nx` and `Ny`. In multi-dimensional problems it makes less sense to specify a Courant number since the wave velocity is a vector and mesh spacings may differ in the various spatial directions. We therefore give Δt explicitly. The signature of the `solver` function is then

```
def solver(I, V, f, c, Lx, Ly, Nx, Ny, dt, T,
           user_action=None, version='scalar'):
```

Key parameters used in the calculations are created as

```
x = linspace(0, Lx, Nx+1)                  # mesh points in x dir
y = linspace(0, Ly, Ny+1)                  # mesh points in y dir
dx = x[1] - x[0]
dy = y[1] - y[0]
Nt = int(round(T/float(dt)))
t = linspace(0, N*dt, N+1)                 # mesh points in time
Cx2 = (c*dt/dx)**2;  Cy2 = (c*dt/dy)**2    # help variables
dt2 = dt**2
```

Solution arrays We store $u_{i,j}^{n+1}$, $u_{i,j}^{n}$, and $u_{i,j}^{n-1}$ in three two-dimensional arrays,

```
u   = zeros((Nx+1,Ny+1))    # solution array
u_n = [zeros((Nx+1,Ny+1)), zeros((Nx+1,Ny+1))]  # t-dt, t-2*dt
```

where $u_{i,j}^{n+1}$ corresponds to `u[i,j]`, $u_{i,j}^{n}$ to `u_n[i,j]`, and $u_{i,j}^{n-1}$ to `u_nm1[i,j]`.

Index sets It is also convenient to introduce the index sets (cf. Sect. 2.6.4)

```
Ix = range(0, u.shape[0])
Iy = range(0, u.shape[1])
It = range(0, t.shape[0])
```

Computing the solution Inserting the initial condition `I` in `u_n` and making a callback to the user in terms of the `user_action` function is a straightforward generalization of the 1D code from Sect. 2.1.6:

```
for i in Ix:
    for j in Iy:
        u_n[i,j] = I(x[i], y[j])

if user_action is not None:
    user_action(u_n, x, xv, y, yv, t, 0)
```

The `user_action` function has additional arguments compared to the 1D case. The arguments `xv` and `yv` will be commented upon in Sect. 2.12.2.

The key finite difference formula (2.110) for updating the solution at a time level is implemented in a separate function as

```
def advance_scalar(u, u_n, u_nm1, f, x, y, t, n, Cx2, Cy2, dt2,
                   V=None, step1=False):
    Ix = range(0, u.shape[0]);  Iy = range(0, u.shape[1])
    if step1:
        dt = sqrt(dt2)  # save
        Cx2 = 0.5*Cx2;  Cy2 = 0.5*Cy2; dt2 = 0.5*dt2  # redefine
        D1 = 1;    D2 = 0
    else:
        D1 = 2;    D2 = 1
    for i in Ix[1:-1]:
        for j in Iy[1:-1]:
            u_xx = u_n[i-1,j] - 2*u_n[i,j] + u_n[i+1,j]
            u_yy = u_n[i,j-1] - 2*u_n[i,j] + u_n[i,j+1]
            u[i,j] = D1*u_n[i,j] - D2*u_nm1[i,j] + \
                     Cx2*u_xx + Cy2*u_yy + dt2*f(x[i], y[j], t[n])
            if step1:
                u[i,j] += dt*V(x[i], y[j])
```

```
    # Boundary condition u=0
    j = Iy[0]
    for i in Ix: u[i,j] = 0
    j = Iy[-1]
    for i in Ix: u[i,j] = 0
    i = Ix[0]
    for j in Iy: u[i,j] = 0
    i = Ix[-1]
    for j in Iy: u[i,j] = 0
    return u
```

The `step1` variable has been introduced to allow the formula to be reused for the first step, computing $u^1_{i,j}$:

```
u = advance_scalar(u, u_n, f, x, y, t,
                   n, Cx2, Cy2, dt, V, step1=True)
```

Below, we will make many alternative implementations of the `advance_scalar` function to speed up the code since most of the CPU time in simulations is spent in this function.

Remark: How to use the solution

The `solver` function in the `wave2D_u0.py` code updates arrays for the next time step by switching references as described in Sect. 2.4.5. Any use of `u` on the user's side is assumed to take place in the user action function. However, should the code be changed such that `u` is returned and used as solution, have in mind that you must return `u_n` after the time limit, otherwise a `return u` will actually return `u_nm1` (due to the switching of array indices in the loop)!

2.12.2 Vectorized Computations

The scalar code above turns out to be extremely slow for large 2D meshes, and probably useless in 3D beyond debugging of small test cases. Vectorization is therefore a must for multi-dimensional finite difference computations in Python. For example, with a mesh consisting of 30×30 cells, vectorization brings down the CPU time by a factor of 70 (!). Equally important, vectorized code can also easily be parallelized to take (usually) optimal advantage of parallel computer platforms.

In the vectorized case, we must be able to evaluate user-given functions like $I(x, y)$ and $f(x, y, t)$ for the entire mesh in one operation (without loops). These user-given functions are provided as Python functions `I(x,y)` and `f(x,y,t)`, respectively. Having the one-dimensional coordinate arrays `x` and `y` is not sufficient when calling `I` and `f` in a vectorized way. We must extend `x` and `y` to their vectorized versions `xv` and `yv`:

```
from numpy import newaxis
xv = x[:,newaxis]
yv = y[newaxis,:]
```

```
# or
xv = x.reshape((x.size, 1))
yv = y.reshape((1, y.size))
```

This is a standard required technique when evaluating functions over a 2D mesh, say `sin(xv)*cos(xv)`, which then gives a result with shape `(Nx+1,Ny+1)`. Calling `I(xv, yv)` and `f(xv, yv, t[n])` will now return `I` and `f` values for the entire set of mesh points.

With the xv and yv arrays for vectorized computing, setting the initial condition is just a matter of

```
u_n[:,:] = I(xv, yv)
```

One could also have written `u_n = I(xv, yv)` and let `u_n` point to a new object, but vectorized operations often make use of direct insertion in the original array through `u_n[:,:]`, because sometimes not all of the array is to be filled by such a function evaluation. This is the case with the computational scheme for $u_{i,j}^{n+1}$:

```
def advance_vectorized(u, u_n, u_nm1, f_a, Cx2, Cy2, dt2,
                       V=None, step1=False):
    if step1:
        dt = np.sqrt(dt2)  # save
        Cx2 = 0.5*Cx2;  Cy2 = 0.5*Cy2; dt2 = 0.5*dt2  # redefine
        D1 = 1;  D2 = 0
    else:
        D1 = 2;  D2 = 1
    u_xx = u_n[:-2,1:-1] - 2*u_n[1:-1,1:-1] + u_n[2:,1:-1]
    u_yy = u_n[1:-1,:-2] - 2*u_n[1:-1,1:-1] + u_n[1:-1,2:]
    u[1:-1,1:-1] = D1*u_n[1:-1,1:-1] - D2*u_nm1[1:-1,1:-1] + \
                   Cx2*u_xx + Cy2*u_yy + dt2*f_a[1:-1,1:-1]
    if step1:
        u[1:-1,1:-1] += dt*V[1:-1, 1:-1]
    # Boundary condition u=0
    j = 0
    u[:,j] = 0
    j = u.shape[1]-1
    u[:,j] = 0
    i = 0
    u[i,:] = 0
    i = u.shape[0]-1
    u[i,:] = 0
    return u
```

Array slices in 2D are more complicated to understand than those in 1D, but the logic from 1D applies to each dimension separately. For example, when doing $u_{i,j}^n - u_{i-1,j}^n$ for $i \in \mathcal{I}_x^+$, we just keep `j` constant and make a slice in the first index: `u_n[1:,j] - u_n[:-1,j]`, exactly as in 1D. The `1:` slice specifies all the indices $i = 1, 2, \ldots, N_x$ (up to the last valid index), while `:-1` specifies the relevant indices for the second term: $0, 1, \ldots, N_x - 1$ (up to, but not including the last index).

In the above code segment, the situation is slightly more complicated, because each displaced slice in one direction is accompanied by a `1:-1` slice in the other direction. The reason is that we only work with the internal points for the index that is kept constant in a difference.

The boundary conditions along the four sides make use of a slice consisting of all indices along a boundary:

```
u[: ,0] = 0
u[:,Ny] = 0
u[0 ,:] = 0
u[Nx,:] = 0
```

In the vectorized update of u (above), the function f is first computed as an array over all mesh points:

```
f_a = f(xv, yv, t[n])
```

We could, alternatively, have used the call `f(xv, yv, t[n])[1:-1,1:-1]` in the last term of the update statement, but other implementations in compiled languages benefit from having f available in an array rather than calling our Python function `f(x,y,t)` for every point.

Also in the `advance_vectorized` function we have introduced a boolean `step1` to reuse the formula for the first time step in the same way as we did with `advance_scalar`. We refer to the `solver` function in `wave2D_u0.py` for the details on how the overall algorithm is implemented.

The callback function now has the arguments `u, x, xv, y, yv, t, n`. The inclusion of xv and yv makes it easy to, e.g., compute an exact 2D solution in the callback function and compute errors, through an expression like `u - u_exact(xv, yv, t[n])`.

2.12.3 Verification

Testing a quadratic solution The 1D solution from Sect. 2.2.4 can be generalized to multi-dimensions and provides a test case where the exact solution also fulfills the discrete equations, such that we know (to machine precision) what numbers the solver function should produce. In 2D we use the following generalization of (2.30):

$$u_e(x,y,t) = x(L_x - x)y(L_y - y)\left(1 + \frac{1}{2}t\right). \tag{2.119}$$

This solution fulfills the PDE problem if $I(x,y) = u_e(x,y,0)$, $V = \frac{1}{2}u_e(x,y,0)$, and $f = 2c^2(1 + \frac{1}{2}t)(y(L_y - y) + x(L_x - x))$. To show that u_e also solves the discrete equations, we start with the general results $[D_tD_t1]^n = 0$, $[D_tD_tt]^n = 0$, and $[D_tD_tt^2] = 2$, and use these to compute

$$\begin{aligned}
[D_xD_xu_e]^n_{i,j} &= \left[y(L_y - y)\left(1 + \frac{1}{2}t\right)D_xD_xx(L_x - x)\right]^n_{i,j} \\
&= y_j(L_y - y_j)\left(1 + \frac{1}{2}t_n\right)(-2).
\end{aligned}$$

A similar calculation must be carried out for the $[D_yD_yu_e]^n_{i,j}$ and $[D_tD_tu_e]^n_{i,j}$ terms. One must also show that the quadratic solution fits the special formula for $u^1_{i,j}$. The details are left as Exercise 2.16. The `test_quadratic` function in the `wave2D_u0.py` program implements this verification as a proper test function for the pytest and nose frameworks.

2.12.4 Visualization

Eventually, we are ready for a real application with our code! Look at the `wave2D_u0.py` and the `gaussian` function. It starts with a Gaussian function to see how it propagates in a square with $u = 0$ on the boundaries:

```
def gaussian(plot_method=2, version='vectorized', save_plot=True):
    """
    Initial Gaussian bell in the middle of the domain.
    plot_method=1 applies mesh function,
    =2 means surf, =3 means Matplotlib, =4 means mayavi,
    =0 means no plot.
    """
    # Clean up plot files
    for name in glob('tmp_*.png'):
        os.remove(name)

    Lx = 10
    Ly = 10
    c = 1.0

    from numpy import exp

    def I(x, y):
        """Gaussian peak at (Lx/2, Ly/2)."""
        return exp(-0.5*(x-Lx/2.0)**2 - 0.5*(y-Ly/2.0)**2)

    def plot_u(u, x, xv, y, yv, t, n):
        """User action function for plotting."""
    ...

    Nx = 40; Ny = 40; T = 20
    dt, cpu = solver(I, None, None, c, Lx, Ly, Nx, Ny, -1, T,
                     user_action=plot_u, version=version)
```

Matplotlib We want to animate a 3D surface in Matplotlib, but this is a really slow process and not recommended, so we consider Matplotlib not an option as long as on-screen animation is desired. One can use the recipes for single shots of u, where it does produce high-quality 3D plots.

Gnuplot Let us look at different ways for visualization. We import SciTools as `st` and can access `st.mesh` and `st.surf` in Matplotlib or Gnuplot, but this is not supported except for the Gnuplot package, where it works really well (Fig. 2.8). Then we choose `plot_method=2` (or less relevant `plot_method=1`) and force the

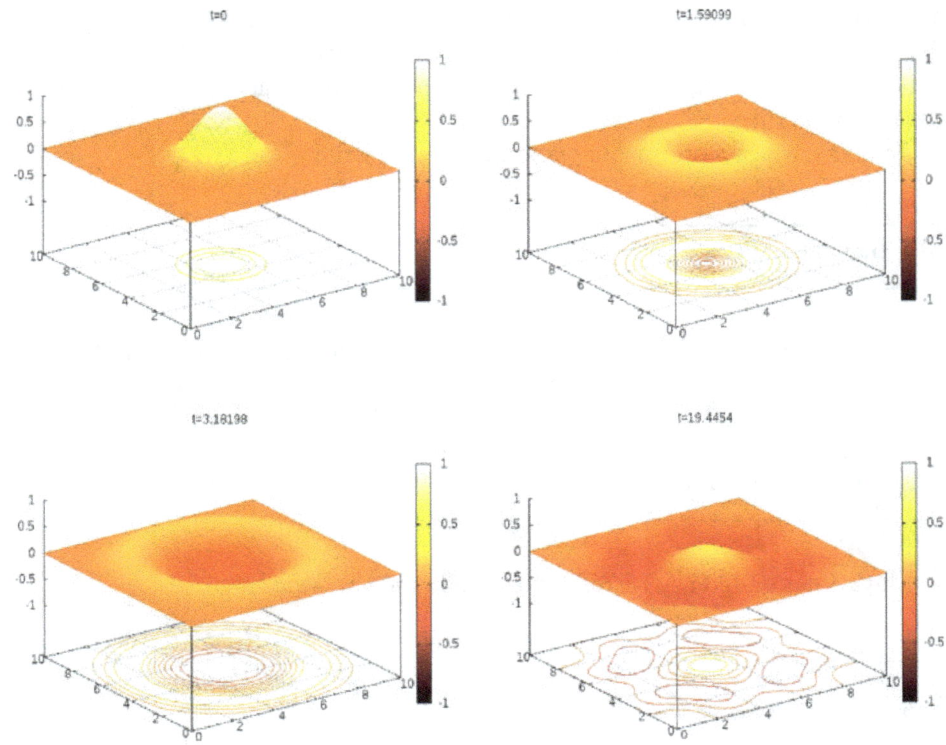

Fig. 2.8 Snapshots of the surface plotted by Gnuplot

backend for SciTools to be Gnuplot (if you have the C package Gnuplot and the `Gnuplot.py` Python interface module installed):

Terminal

```
Terminal> python wave2D_u0.py --SCITOOLS_easyviz_backend gnuplot
```

It gives a nice visualization with lifted surface and contours beneath. Figure 2.8 shows four plots of u.

Video files can be made of the PNG frames:

Terminal

```
Terminal> ffmpeg -i tmp_%04d.png -r 25 -vcodec flv movie.flv
Terminal> ffmpeg -i tmp_%04d.png -r 25 -vcodec linx264 movie.mp4
Terminal> ffmpeg -i tmp_%04d.png -r 25 -vcodec libvpx movie.webm
Terminal> ffmpeg -i tmp_%04d.png -r 25 -vcodec libtheora movie.ogg
```

It is wise to use a high frame rate – a low one will just skip many frames. There may also be considerable quality differences between the different formats.

Movie 1 https://raw.githubusercontent.com/hplgit/fdm-book/master/doc/pub/book/html/mov-wave/gnuplot/wave2D_u0_gaussian/movie25.mp4

Mayavi The best option for doing visualization of 2D and 3D scalar and vector fields in Python programs is Mayavi, which is an interface to the high-quality package VTK in C++. There is good online documentation and also an introduction in Chapter 5 of [10].

To obtain Mayavi on Ubuntu platforms you can write

Terminal
```
pip install mayavi --upgrade
```

For Mac OS X and Windows, we recommend using Anaconda. To obtain Mayavi for Anaconda you can write

Terminal
```
conda install mayavi
```

Mayavi has a MATLAB-like interface called `mlab`. We can do

```
import mayavi.mlab as plt
# or
from mayavi import mlab
```

and have `plt` (as usual) or `mlab` as a kind of MATLAB visualization access inside our program (just more powerful and with higher visual quality).

The official documentation of the `mlab` module is provided in two places, one for the basic functionality[12] and one for further functionality[13]. Basic figure handling[14] is very similar to the one we know from Matplotlib. Just as for Matplotlib, all plotting commands you do in `mlab` will go into the same figure, until you manually change to a new figure.

Back to our application, the following code for the user action function with plotting in Mayavi is relevant to add.

```
# Top of the file
try:
    import mayavi.mlab as mlab
except:
    # We don't have mayavi
    pass

def solver(...):
    ...
```

[12] http://docs.enthought.com/mayavi/mayavi/auto/mlab_helper_functions.html
[13] http://docs.enthought.com/mayavi/mayavi/auto/mlab_other_functions.html
[14] http://docs.enthought.com/mayavi/mayavi/auto/mlab_figure.html

```
def gaussian(...):
    ...
    if plot_method == 3:
        from mpl_toolkits.mplot3d import axes3d
        import matplotlib.pyplot as plt
        from matplotlib import cm
        plt.ion()
        fig = plt.figure()
        u_surf = None

    def plot_u(u, x, xv, y, yv, t, n):
        """User action function for plotting."""
        if t[n] == 0:
            time.sleep(2)
        if plot_method == 1:
            # Works well with Gnuplot backend, not with Matplotlib
            st.mesh(x, y, u, title='t=%g' % t[n], zlim=[-1,1],
                    caxis=[-1,1])
        elif plot_method == 2:
            # Works well with Gnuplot backend, not with Matplotlib
            st.surfc(xv, yv, u, title='t=%g' % t[n], zlim=[-1, 1],
                  colorbar=True, colormap=st.hot(), caxis=[-1,1],
                  shading='flat')
        elif plot_method == 3:
            print 'Experimental 3D matplotlib...not recommended'
        elif plot_method == 4:
        # Mayavi visualization
            mlab.clf()
            extent1 = (0, 20, 0, 20,-2, 2)
            s = mlab.surf(x , y, u,
                          colormap='Blues',
                          warp_scale=5,extent=extent1)
            mlab.axes(s, color=(.7, .7, .7), extent=extent1,
                      ranges=(0, 10, 0, 10, -1, 1),
                      xlabel='', ylabel='', zlabel='',
                      x_axis_visibility=False,
                      z_axis_visibility=False)
            mlab.outline(s, color=(0.7, .7, .7), extent=extent1)
            mlab.text(6, -2.5, '', z=-4, width=0.14)
            mlab.colorbar(object=None, title=None,
                          orientation='horizontal',
                          nb_labels=None, nb_colors=None,
                          label_fmt=None)
            mlab.title('Gaussian t=%g' % t[n])
            mlab.view(142, -72, 50)
            f = mlab.gcf()
            camera = f.scene.camera
            camera.yaw(0)


        if plot_method > 0:
            time.sleep(0) # pause between frames
            if save_plot:
                filename = 'tmp_%04d.png' % n
        if plot_method == 4:
                  mlab.savefig(filename)  # time consuming!
        elif plot_method in (1,2):
                  st.savefig(filename)  # time consuming!
```

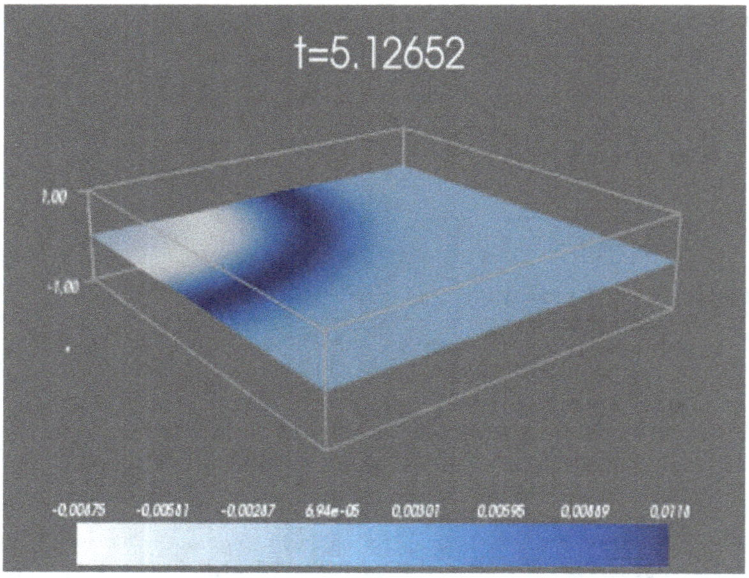

Fig. 2.9 Plot with Mayavi

This is a point to get started – visualization is as always a very time-consuming and experimental discipline. With the PNG files we can use `ffmpeg` to create videos.

Movie 2 https://github.com/hplgit/fdm-book/blob/master/doc/pub/book/html/mov-wave/mayavi/wave2D_u0_gaussian/movie.mp4

2.13 Exercises

Exercise 2.16: Check that a solution fulfills the discrete model
Carry out all mathematical details to show that (2.119) is indeed a solution of the discrete model for a 2D wave equation with $u = 0$ on the boundary. One must check the boundary conditions, the initial conditions, the general discrete equation at a time level and the special version of this equation for the first time level.
Filename: `check_quadratic_solution`.

Project 2.17: Calculus with 2D mesh functions
The goal of this project is to redo Project 2.6 with 2D mesh functions $(f_{i,j})$.

Differentiation The differentiation results in a discrete gradient function, which in the 2D case can be represented by a three-dimensional array `df[d,i,j]` where `d` represents the direction of the derivative, and `i,j` is a mesh point in 2D. Use centered differences for the derivative at inner points and one-sided forward or backward differences at the boundary points. Construct unit tests and write a corresponding test function.

Integration The integral of a 2D mesh function $f_{i,j}$ is defined as

$$F_{i,j} = \int_{y_0}^{y_j}\int_{x_0}^{x_i} f(x,y)dxdy,$$

where $f(x,y)$ is a function that takes on the values of the discrete mesh function $f_{i,j}$ at the mesh points, but can also be evaluated in between the mesh points. The particular variation between mesh points can be taken as bilinear, but this is not important as we will use a product Trapezoidal rule to approximate the integral over a cell in the mesh and then we only need to evaluate $f(x,y)$ at the mesh points.

Suppose $F_{i,j}$ is computed. The calculation of $F_{i+1,j}$ is then

$$\begin{aligned} F_{i+1,j} &= F_{i,j} + \int_{x_i}^{x_{i+1}}\int_{y_0}^{y_j} f(x,y)dydx \\ &\approx \Delta x\frac{1}{2}\left(\int_{y_0}^{y_j} f(x_i,y)dy + \int_{y_0}^{y_j} f(x_{i+1},y)dy\right). \end{aligned}$$

The integrals in the y direction can be approximated by a Trapezoidal rule. A similar idea can be used to compute $F_{i,j+1}$. Thereafter, $F_{i+1,j+1}$ can be computed by adding the integral over the final corner cell to $F_{i+1,j} + F_{i,j+1} - F_{i,j}$. Carry out the details of these computations and implement a function that can return $F_{i,j}$ for all mesh indices i and j. Use the fact that the Trapezoidal rule is exact for linear functions and write a test function.
Filename: `mesh_calculus_2D`.

Exercise 2.18: Implement Neumann conditions in 2D
Modify the `wave2D_u0.py` program, which solves the 2D wave equation $u_{tt} = c^2(u_{xx} + u_{yy})$ with constant wave velocity c and $u = 0$ on the boundary, to have Neumann boundary conditions: $\partial u/\partial n = 0$. Include both scalar code (for debugging and reference) and vectorized code (for speed).

To test the code, use $u = 1.2$ as solution ($I(x,y) = 1.2$, $V = f = 0$, and c arbitrary), which should be exactly reproduced with any mesh as long as the stability criterion is satisfied. Another test is to use the plug-shaped pulse in the `pulse` function from Sect. 2.8 and the `wave1D_dn_vc.py` program. This pulse is exactly propagated in 1D if $c\Delta t/\Delta x = 1$. Check that also the 2D program can propagate this pulse exactly in x direction ($c\Delta t/\Delta x = 1$, Δy arbitrary) and y direction ($c\Delta t/\Delta y = 1$, Δx arbitrary).
Filename: `wave2D_dn`.

Exercise 2.19: Test the efficiency of compiled loops in 3D
Extend the `wave2D_u0.py` code and the Cython, Fortran, and C versions to 3D. Set up an efficiency experiment to determine the relative efficiency of pure scalar Python code, vectorized code, Cython-compiled loops, Fortran-compiled loops, and C-compiled loops. Normalize the CPU time for each mesh by the fastest version.
Filename: `wave3D_u0`.

2.14 Applications of Wave Equations

This section presents a range of wave equation models for different physical phenomena. Although many wave motion problems in physics can be modeled by the standard linear wave equation, or a similar formulation with a system of first-order equations, there are some exceptions. Perhaps the most important is water waves: these are modeled by the Laplace equation with time-dependent boundary conditions at the water surface (long water waves, however, can be approximated by a standard wave equation, see Sect. 2.14.7). Quantum mechanical waves constitute another example where the waves are governed by the Schrödinger equation, i.e., not by a standard wave equation. Many wave phenomena also need to take nonlinear effects into account when the wave amplitude is significant. Shock waves in the air is a primary example.

The derivations in the following are very brief. Those with a firm background in continuum mechanics will probably have enough knowledge to fill in the details, while other readers will hopefully get some impression of the physics and approximations involved when establishing wave equation models.

2.14.1 Waves on a String

Figure 2.10 shows a model we may use to derive the equation for waves on a string. The string is modeled as a set of discrete point masses (at mesh points) with elastic strings in between. The string has a large constant tension T. We let the mass at mesh point x_i be m_i. The displacement of this mass point in the y direction is denoted by $u_i(t)$.

The motion of mass m_i is governed by Newton's second law of motion. The position of the mass at time t is $x_i\boldsymbol{i} + u_i(t)\boldsymbol{j}$, where $\boldsymbol{i}$ and $\boldsymbol{j}$ are unit vectors in the x and y direction, respectively. The acceleration is then $u_i''(t)\boldsymbol{j}$. Two forces are

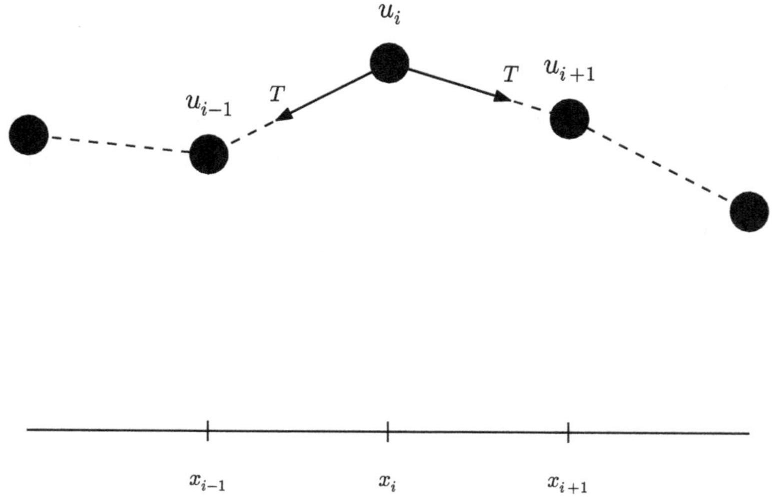

Fig. 2.10 Discrete string model with point masses connected by elastic strings

acting on the mass as indicated in Fig. 2.10. The force $\boldsymbol{T}^-$ acting toward the point x_{i-1} can be decomposed as

$$\boldsymbol{T}^- = -T\sin\phi\boldsymbol{i} - T\cos\phi\boldsymbol{j},$$

where ϕ is the angle between the force and the line $x = x_i$. Let $\Delta u_i = u_i - u_{i-1}$ and let $\Delta s_i = \sqrt{\Delta u_i^2 + (x_i - x_{i-1})^2}$ be the distance from mass m_{i-1} to mass m_i. It is seen that $\cos\phi = \Delta u_i/\Delta s_i$ and $\sin\phi = (x_i - x_{i-1})/\Delta s$ or $\Delta x/\Delta s_i$ if we introduce a constant mesh spacing $\Delta x = x_i - x_{i-1}$. The force can then be written

$$\boldsymbol{T}^- = -T\frac{\Delta x}{\Delta s_i}\boldsymbol{i} - T\frac{\Delta u_i}{\Delta s_i}\boldsymbol{j}.$$

The force $\boldsymbol{T}^+$ acting toward x_{i+1} can be calculated in a similar way:

$$\boldsymbol{T}^+ = T\frac{\Delta x}{\Delta s_{i+1}}\boldsymbol{i} + T\frac{\Delta u_{i+1}}{\Delta s_{i+1}}\boldsymbol{j}.$$

Newton's second law becomes

$$m_i u_i''(t)\boldsymbol{j} = \boldsymbol{T}^+ + \boldsymbol{T}^-,$$

which gives the component equations

$$T\frac{\Delta x}{\Delta s_i} = T\frac{\Delta x}{\Delta s_{i+1}}, \tag{2.120}$$

$$m_i u_i''(t) = T\frac{\Delta u_{i+1}}{\Delta s_{i+1}} - T\frac{\Delta u_i}{\Delta s_i}. \tag{2.121}$$

A basic reasonable assumption for a string is small displacements u_i and small displacement gradients $\Delta u_i/\Delta x$. For small $g = \Delta u_i/\Delta x$ we have that

$$\Delta s_i = \sqrt{\Delta u_i^2 + \Delta x^2} = \Delta x\sqrt{1+g^2} + \Delta x\left(1 + \frac{1}{2}g^2 + \mathcal{O}(g^4)\right) \approx \Delta x.$$

Equation (2.120) is then simply the identity $T = T$, while (2.121) can be written as

$$m_i u_i''(t) = T\frac{\Delta u_{i+1}}{\Delta x} - T\frac{\Delta u_i}{\Delta x},$$

which upon division by Δx and introducing the density $\varrho_i = m_i/\Delta x$ becomes

$$\varrho_i u_i''(t) = T\frac{1}{\Delta x^2}\left(u_{i+1} - 2u_i + u_{i-1}\right). \tag{2.122}$$

We can now choose to approximate u_i'' by a finite difference in time and get the discretized wave equation,

$$\varrho_i\frac{1}{\Delta t^2}\left(u_i^{n+1} - 2u_i^n - u_i^{n-1}\right) = T\frac{1}{\Delta x^2}\left(u_{i+1} - 2u_i + u_{i-1}\right). \tag{2.123}$$

On the other hand, we may go to the continuum limit $\Delta x \to 0$ and replace $u_i(t)$ by $u(x,t)$, ϱ_i by $\varrho(x)$, and recognize that the right-hand side of (2.122) approaches $\partial^2 u/\partial x^2$ as $\Delta x \to 0$. We end up with the continuous model for waves on a string:

$$\varrho\frac{\partial^2 u}{\partial t^2} = T\frac{\partial^2 u}{\partial x^2}\,. \tag{2.124}$$

Note that the density ϱ may change along the string, while the tension T is a constant. With variable wave velocity $c(x) = \sqrt{T/\varrho(x)}$ we can write the wave equation in the more standard form

$$\frac{\partial^2 u}{\partial t^2} = c^2(x)\frac{\partial^2 u}{\partial x^2}\,. \tag{2.125}$$

Because of the way ϱ enters the equations, the variable wave velocity does *not* appear inside the derivatives as in many other versions of the wave equation. However, most strings of interest have constant ϱ.

The end points of a string are fixed so that the displacement u is zero. The boundary conditions are therefore $u = 0$.

Damping Air resistance and non-elastic effects in the string will contribute to reduce the amplitudes of the waves so that the motion dies out after some time. This damping effect can be modeled by a term bu_t on the left-hand side of the equation

$$\varrho\frac{\partial^2 u}{\partial t^2} + b\frac{\partial u}{\partial t} = T\frac{\partial^2 u}{\partial x^2}\,. \tag{2.126}$$

The parameter $b \geq 0$ is small for most wave phenomena, but the damping effect may become significant in long time simulations.

External forcing It is easy to include an external force acting on the string. Say we have a vertical force $\tilde{f}_i\boldsymbol{j}$ acting on mass m_i, modeling the effect of gravity on a string. This force affects the vertical component of Newton's law and gives rise to an extra term $\tilde{f}(x,t)$ on the right-hand side of (2.124). In the model (2.125) we would add a term $f(x,t) = \tilde{f}(x,t)/\varrho(x)$.

Modeling the tension via springs We assumed, in the derivation above, that the tension in the string, T, was constant. It is easy to check this assumption by modeling the string segments between the masses as standard springs, where the force (tension T) is proportional to the elongation of the spring segment. Let k be the spring constant, and set $T_i = k\Delta\ell$ for the tension in the spring segment between x_{i-1} and x_i, where $\Delta\ell$ is the elongation of this segment from the tension-free state. A basic feature of a string is that it has high tension in the equilibrium position $u = 0$. Let the string segment have an elongation $\Delta\ell_0$ in the equilibrium position. After deformation of the string, the elongation is $\Delta\ell = \Delta\ell_0 + \Delta s_i$: $T_i = k(\Delta\ell_0 + \Delta s_i) \approx k(\Delta\ell_0 + \Delta x)$. This shows that T_i is independent of i. Moreover, the extra approximate elongation Δx is very small compared to $\Delta\ell_0$, so we may well set $T_i = T = k\Delta\ell_0$. This means that the tension is completely dominated by the initial tension determined by the tuning of the string. The additional deformations of the spring during the vibrations do not introduce significant changes in the tension.

2.14.2 Elastic Waves in a Rod

Consider an elastic rod subject to a hammer impact at the end. This experiment will give rise to an elastic deformation pulse that travels through the rod. A mathematical model for longitudinal waves along an elastic rod starts with the general equation for deformations and stresses in an elastic medium,

$$\varrho \boldsymbol{u}_{tt} = \nabla \cdot \boldsymbol{\sigma} + \varrho \boldsymbol{f}, \tag{2.127}$$

where ϱ is the density, $\boldsymbol{u}$ the displacement field, $\boldsymbol{\sigma}$ the stress tensor, and $\boldsymbol{f}$ body forces. The latter has normally no impact on elastic waves.

For stationary deformation of an elastic rod, aligned with the x axis, one has that $\sigma_{xx} = Eu_x$, with all other stress components being zero. The parameter E is known as Young's modulus. Moreover, we set $\boldsymbol{u} = u(x,t)\boldsymbol{i}$ and neglect the radial contraction and expansion (where Poisson's ratio is the important parameter). Assuming that this simple stress and deformation field is a good approximation, (2.127) simplifies to

$$\varrho \frac{\partial^2 u}{\partial t^2} = \frac{\partial}{\partial x}\left(E\frac{\partial u}{\partial x}\right). \tag{2.128}$$

The associated boundary conditions are u or $\sigma_{xx} = Eu_x$ known, typically $u = 0$ for a fixed end and $\sigma_{xx} = 0$ for a free end.

2.14.3 Waves on a Membrane

Think of a thin, elastic membrane with shape as a circle or rectangle. This membrane can be brought into oscillatory motion and will develop elastic waves. We can model this phenomenon somewhat similar to waves in a rod: waves in a membrane are simply the two-dimensional counterpart. We assume that the material is deformed in the z direction only and write the elastic displacement field on the form $\boldsymbol{u}(x,y,t) = w(x,y,t)\boldsymbol{i}$. The z coordinate is omitted since the membrane is thin and all properties are taken as constant throughout the thickness. Inserting this displacement field in Newton's 2nd law of motion (2.127) results in

$$\varrho \frac{\partial^2 w}{\partial t^2} = \frac{\partial}{\partial x}\left(\mu\frac{\partial w}{\partial x}\right) + \frac{\partial}{\partial y}\left(\mu\frac{\partial w}{\partial y}\right). \tag{2.129}$$

This is nothing but a wave equation in $w(x,y,t)$, which needs the usual initial conditions on w and w_t as well as a boundary condition $w = 0$. When computing the stress in the membrane, one needs to split $\boldsymbol{\sigma}$ into a constant high-stress component due to the fact that all membranes are normally pre-stressed, plus a component proportional to the displacement and governed by the wave motion.

2.14.4 The Acoustic Model for Seismic Waves

Seismic waves are used to infer properties of subsurface geological structures. The physical model is a heterogeneous elastic medium where sound is propagated by small elastic vibrations. The general mathematical model for deformations in an elastic medium is based on Newton's second law,

$$\varrho \boldsymbol{u}_{tt} = \nabla \cdot \boldsymbol{\sigma} + \varrho \boldsymbol{f}, \tag{2.130}$$

and a constitutive law relating $\boldsymbol{\sigma}$ to $\boldsymbol{u}$, often Hooke's generalized law,

$$\boldsymbol{\sigma} = K\nabla \cdot \boldsymbol{u}\,\boldsymbol{I} + G\left(\nabla \boldsymbol{u} + (\nabla \boldsymbol{u})^T - \frac{2}{3}\nabla \cdot \boldsymbol{u}\,\boldsymbol{I}\right). \tag{2.131}$$

Here, $\boldsymbol{u}$ is the displacement field, $\boldsymbol{\sigma}$ is the stress tensor, $\boldsymbol{I}$ is the identity tensor, ϱ is the medium's density, $\boldsymbol{f}$ are body forces (such as gravity), K is the medium's bulk modulus and G is the shear modulus. All these quantities may vary in space, while $\boldsymbol{u}$ and $\boldsymbol{\sigma}$ will also show significant variation in time during wave motion.

The acoustic approximation to elastic waves arises from a basic assumption that the second term in Hooke's law, representing the deformations that give rise to shear stresses, can be neglected. This assumption can be interpreted as approximating the geological medium by a fluid. Neglecting also the body forces $\boldsymbol{f}$, (2.130) becomes

$$\varrho \boldsymbol{u}_{tt} = \nabla(K\nabla \cdot \boldsymbol{u}). \tag{2.132}$$

Introducing p as a pressure via

$$p = -K\nabla \cdot \boldsymbol{u}, \tag{2.133}$$

and dividing (2.132) by ϱ, we get

$$\boldsymbol{u}_{tt} = -\frac{1}{\varrho}\nabla p. \tag{2.134}$$

Taking the divergence of this equation, using $\nabla \cdot \boldsymbol{u} = -p/K$ from (2.133), gives the *acoustic approximation to elastic waves*:

$$p_{tt} = K\nabla \cdot \left(\frac{1}{\varrho}\nabla p\right). \tag{2.135}$$

This is a standard, linear wave equation with variable coefficients. It is common to add a source term $s(x, y, z, t)$ to model the generation of sound waves:

$$p_{tt} = K\nabla \cdot \left(\frac{1}{\varrho}\nabla p\right) + s. \tag{2.136}$$

A common additional approximation of (2.136) is based on using the chain rule on the right-hand side,

$$K\nabla \cdot \left(\frac{1}{\varrho}\nabla p\right) = \frac{K}{\varrho}\nabla^2 p + K\nabla\left(\frac{1}{\varrho}\right) \cdot \nabla p \approx \frac{K}{\varrho}\nabla^2 p,$$

under the assumption that the relative spatial gradient $\nabla\varrho^{-1} = -\varrho^{-2}\nabla\varrho$ is small. This approximation results in the simplified equation

$$p_{tt} = \frac{K}{\varrho}\nabla^2 p + s\,. \tag{2.137}$$

The acoustic approximations to seismic waves are used for sound waves in the ground, and the Earth's surface is then a boundary where p equals the atmospheric pressure p_0 such that the boundary condition becomes $p = p_0$.

Anisotropy Quite often in geological materials, the effective wave velocity $c = \sqrt{K/\varrho}$ is different in different spatial directions because geological layers are compacted, and often twisted, in such a way that the properties in the horizontal and vertical direction differ. With z as the vertical coordinate, we can introduce a vertical wave velocity c_z and a horizontal wave velocity c_h, and generalize (2.137) to

$$p_{tt} = c_z^2 p_{zz} + c_h^2(p_{xx} + p_{yy}) + s\,. \tag{2.138}$$

2.14.5 Sound Waves in Liquids and Gases

Sound waves arise from pressure and density variations in fluids. The starting point of modeling sound waves is the basic equations for a compressible fluid where we omit viscous (frictional) forces, body forces (gravity, for instance), and temperature effects:

$$\varrho_t + \nabla\cdot(\varrho\boldsymbol{u}) = 0, \tag{2.139}$$

$$\varrho\boldsymbol{u}_t + \varrho\boldsymbol{u}\cdot\nabla\boldsymbol{u} = -\nabla p, \tag{2.140}$$

$$\varrho = \varrho(p)\,. \tag{2.141}$$

These equations are often referred to as the Euler equations for the motion of a fluid. The parameters involved are the density ϱ, the velocity $\boldsymbol{u}$, and the pressure p. Equation (2.139) reflects mass balance, (2.140) is Newton's second law for a fluid, with frictional and body forces omitted, and (2.141) is a constitutive law relating density to pressure by thermodynamic considerations. A typical model for (2.141) is the so-called isentropic relation[15], valid for adiabatic processes where there is no heat transfer:

$$\varrho = \varrho_0\left(\frac{p}{p_0}\right)^{1/\gamma}. \tag{2.142}$$

Here, p_0 and ϱ_0 are reference values for p and ϱ when the fluid is at rest, and γ is the ratio of specific heat at constant pressure and constant volume ($\gamma = 5/3$ for air).

The key approximation in a mathematical model for sound waves is to assume that these waves are small perturbations to the density, pressure, and velocity. We

[15] http://en.wikipedia.org/wiki/Isentropic_process

therefore write

$$
\begin{aligned}
p &= p_0 + \hat{p},\\
\varrho &= \varrho_0 + \hat{\varrho},\\
\boldsymbol{u} &= \hat{\boldsymbol{u}},
\end{aligned}
$$

where we have decomposed the fields in a constant equilibrium value, corresponding to $\boldsymbol{u} = 0$, and a small perturbation marked with a hat symbol. By inserting these decompositions in (2.139) and (2.140), neglecting all product terms of small perturbations and/or their derivatives, and dropping the hat symbols, one gets the following linearized PDE system for the small perturbations in density, pressure, and velocity:

$$
\varrho_t + \varrho_0 \nabla\cdot\boldsymbol{u} = 0, \tag{2.143}
$$

$$
\varrho_0 \boldsymbol{u}_t = -\nabla p\,. \tag{2.144}
$$

Now we can eliminate ϱ_t by differentiating the relation $\varrho(p)$,

$$
\varrho_t = \varrho_0 \frac{1}{\gamma}\left(\frac{p}{p_0}\right)^{1/\gamma - 1}\frac{1}{p_0}p_t = \frac{\varrho_0}{\gamma p_0}\left(\frac{p}{p_0}\right)^{1/\gamma-1} p_t\,.
$$

The product term $p^{1/\gamma-1}p_t$ can be linearized as $p_0^{1/\gamma-1}p_t$, resulting in

$$
\varrho_t \approx \frac{\varrho_0}{\gamma p_0} p_t\,.
$$

We then get

$$
p_t + \gamma p_0 \nabla\cdot\boldsymbol{u} = 0, \tag{2.145}
$$

$$
\boldsymbol{u}_t = -\frac{1}{\varrho_0}\nabla p\,. \tag{2.146}
$$

Taking the divergence of (2.146) and differentiating (2.145) with respect to time gives the possibility to easily eliminate $\nabla\cdot\boldsymbol{u}_t$ and arrive at a standard, linear wave equation for p:

$$
p_{tt} = c^2\nabla^2 p, \tag{2.147}
$$

where $c = \sqrt{\gamma p_0/\varrho_0}$ is the speed of sound in the fluid.

2.14.6 Spherical Waves

Spherically symmetric three-dimensional waves propagate in the radial direction r only so that $u = u(r,t)$. The fully three-dimensional wave equation

$$
\frac{\partial^2 u}{\partial t^2} = \nabla\cdot(c^2\nabla u) + f
$$

then reduces to the spherically symmetric wave equation

$$
\frac{\partial^2 u}{\partial t^2} = \frac{1}{r^2}\frac{\partial}{\partial r}\left(c^2(r)r^2\frac{\partial u}{\partial r}\right) + f(r,t), \quad r\in(0,R),\ t>0\,. \tag{2.148}
$$

One can easily show that the function $v(r,t) = ru(r,t)$ fulfills a standard wave equation in Cartesian coordinates if c is constant. To this end, insert $u = v/r$ in

$$\frac{1}{r^2}\frac{\partial}{\partial r}\left(c^2(r)r^2\frac{\partial u}{\partial r}\right)$$

to obtain

$$r\left(\frac{dc^2}{dr}\frac{\partial v}{\partial r} + c^2\frac{\partial^2 v}{\partial r^2}\right) - \frac{dc^2}{dr}v\,.$$

The two terms in the parenthesis can be combined to

$$r\frac{\partial}{\partial r}\left(c^2\frac{\partial v}{\partial r}\right),$$

which is recognized as the variable-coefficient Laplace operator in one Cartesian coordinate. The spherically symmetric wave equation in terms of $v(r,t)$ now becomes

$$\frac{\partial^2 v}{\partial t^2} = \frac{\partial}{\partial r}\left(c^2(r)\frac{\partial v}{\partial r}\right) - \frac{1}{r}\frac{dc^2}{dr}v + rf(r,t), \quad r \in (0,R),\ t > 0\,. \tag{2.149}$$

In the case of constant wave velocity c, this equation reduces to the wave equation in a single Cartesian coordinate called r:

$$\frac{\partial^2 v}{\partial t^2} = c^2\frac{\partial^2 v}{\partial r^2} + rf(r,t), \quad r \in (0,R),\ t > 0\,. \tag{2.150}$$

That is, any program for solving the one-dimensional wave equation in a Cartesian coordinate system can be used to solve (2.150), provided the source term is multiplied by the coordinate, and that we divide the Cartesian mesh solution by r to get the spherically symmetric solution. Moreover, if $r = 0$ is included in the domain, spherical symmetry demands that $\partial u/\partial r = 0$ at $r = 0$, which means that

$$\frac{\partial u}{\partial r} = \frac{1}{r^2}\left(r\frac{\partial v}{\partial r} - v\right) = 0, \quad r = 0\,.$$

For this to hold in the limit $r \to 0$, we must have $v(0,t) = 0$ at least as a necessary condition. In most practical applications, we exclude $r = 0$ from the domain and assume that some boundary condition is assigned at $r = \epsilon$, for some $\epsilon > 0$.

2.14.7 The Linear Shallow Water Equations

The next example considers water waves whose wavelengths are much larger than the depth and whose wave amplitudes are small. This class of waves may be generated by catastrophic geophysical events, such as earthquakes at the sea bottom, landslides moving into water, or underwater slides (or a combination, as earthquakes frequently release avalanches of masses). For example, a subsea earthquake will normally have an extension of many kilometers but lift the water only a few meters. The wave length will have a size dictated by the earthquake area, which is much lager than the water depth, and compared to this wave length, an amplitude of a few meters is very small. The water is essentially a thin film, and mathematically we can average the problem in the vertical direction and approximate the 3D wave phenomenon by 2D PDEs. Instead of a moving water domain in three space dimensions, we get a horizontal 2D domain with an unknown function for the surface elevation and the water depth as a variable coefficient in the PDEs.

Let $\eta(x, y, t)$ be the elevation of the water surface, $H(x, y)$ the water depth corresponding to a flat surface ($\eta = 0$), $u(x, y, t)$ and $v(x, y, t)$ the depth-averaged horizontal velocities of the water. Mass and momentum balance of the water volume give rise to the PDEs involving these quantities:

$$\eta_t = -(Hu)_x - (Hv)_x \tag{2.151}$$

$$u_t = -g\eta_x, \tag{2.152}$$

$$v_t = -g\eta_y, \tag{2.153}$$

where g is the acceleration of gravity. Equation (2.151) corresponds to mass balance while the other two are derived from momentum balance (Newton's second law).

The initial conditions associated with (2.151)–(2.153) are η, u, and v prescribed at $t = 0$. A common condition is to have some water elevation $\eta = I(x, y)$ and assume that the surface is at rest: $u = v = 0$. A subsea earthquake usually means a sufficiently rapid motion of the bottom and the water volume to say that the bottom deformation is mirrored at the water surface as an initial lift $I(x, y)$ and that $u = v = 0$.

Boundary conditions may be η prescribed for incoming, known waves, or zero normal velocity at reflecting boundaries (steep mountains, for instance): $un_x + vn_y = 0$, where (n_x, n_y) is the outward unit normal to the boundary. More sophisticated boundary conditions are needed when waves run up at the shore, and at open boundaries where we want the waves to leave the computational domain undisturbed.

Equations (2.151), (2.152), and (2.153) can be transformed to a standard, linear wave equation. First, multiply (2.152) and (2.153) by H, differentiate (2.152) with respect to x and (2.153) with respect to y. Second, differentiate (2.151) with respect to t and use that $(Hu)_{xt} = (Hu_t)_x$ and $(Hv)_{yt} = (Hv_t)_y$ when H is independent of t. Third, eliminate $(Hu_t)_x$ and $(Hv_t)_y$ with the aid of the other two differentiated equations. These manipulations result in a standard, linear wave equation for η:

$$\eta_{tt} = (gH\eta_x)_x + (gH\eta_y)_y = \nabla\cdot(gH\nabla\eta)\,. \tag{2.154}$$

In the case we have an initial non-flat water surface at rest, the initial conditions become $\eta = I(x, y)$ and $\eta_t = 0$. The latter follows from (2.151) if $u = v = 0$, or simply from the fact that the vertical velocity of the surface is η_t, which is zero for a surface at rest.

The system (2.151)–(2.153) can be extended to handle a time-varying bottom topography, which is relevant for modeling long waves generated by underwater slides. In such cases the water depth function H is also a function of t, due to the moving slide, and one must add a time-derivative term H_t to the left-hand side of

(2.151). A moving bottom is best described by introducing $z = H_0$ as the still-water level, $z = B(x, y, t)$ as the time- and space-varying bottom topography, so that $H = H_0 - B(x, y, t)$. In the elimination of u and v one may assume that the dependence of H on t can be neglected in the terms $(Hu)_{xt}$ and $(Hv)_{yt}$. We then end up with a source term in (2.154), because of the moving (accelerating) bottom:

$$\eta_{tt} = \nabla \cdot (gH\nabla\eta) + B_{tt}\,. \tag{2.155}$$

The reduction of (2.155) to 1D, for long waves in a straight channel, or for approximately plane waves in the ocean, is trivial by assuming no change in y direction ($\partial/\partial y = 0$):

$$\eta_{tt} = (gH\eta_x)_x + B_{tt}\,. \tag{2.156}$$

Wind drag on the surface Surface waves are influenced by the drag of the wind, and if the wind velocity some meters above the surface is (U, V), the wind drag gives contributions $C_V\sqrt{U^2 + V^2}U$ and $C_V\sqrt{U^2 + V^2}V$ to (2.152) and (2.153), respectively, on the right-hand sides.

Bottom drag The waves will experience a drag from the bottom, often roughly modeled by a term similar to the wind drag: $C_B\sqrt{u^2 + v^2}u$ on the right-hand side of (2.152) and $C_B\sqrt{u^2 + v^2}v$ on the right-hand side of (2.153). Note that in this case the PDEs (2.152) and (2.153) become nonlinear and the elimination of u and v to arrive at a 2nd-order wave equation for η is not possible anymore.

Effect of the Earth's rotation Long geophysical waves will often be affected by the rotation of the Earth because of the Coriolis force. This force gives rise to a term fv on the right-hand side of (2.152) and $-fu$ on the right-hand side of (2.153). Also in this case one cannot eliminate u and v to work with a single equation for η. The Coriolis parameter is $f = 2\Omega \sin\phi$, where Ω is the angular velocity of the earth and ϕ is the latitude.

2.14.8 Waves in Blood Vessels

The flow of blood in our bodies is basically fluid flow in a network of pipes. Unlike rigid pipes, the walls in the blood vessels are elastic and will increase their diameter when the pressure rises. The elastic forces will then push the wall back and accelerate the fluid. This interaction between the flow of blood and the deformation of the vessel wall results in waves traveling along our blood vessels.

A model for one-dimensional waves along blood vessels can be derived from averaging the fluid flow over the cross section of the blood vessels. Let x be a coordinate along the blood vessel and assume that all cross sections are circular, though with different radii $R(x, t)$. The main quantities to compute is the cross section area $A(x, t)$, the averaged pressure $P(x, t)$, and the total volume flux $Q(x, t)$. The area of this cross section is

$$A(x, t) = 2\pi \int_0^{R(x,t)} r dr\,. \tag{2.157}$$

Let $v_x(x,t)$ be the velocity of blood averaged over the cross section at point x. The volume flux, being the total volume of blood passing a cross section per time unit, becomes

$$Q(x,t) = A(x,t)v_x(x,t)\,. \tag{2.158}$$

Mass balance and Newton's second law lead to the PDEs

$$\frac{\partial A}{\partial t} + \frac{\partial Q}{\partial x} = 0, \tag{2.159}$$

$$\frac{\partial Q}{\partial t} + \frac{\gamma+2}{\gamma+1}\frac{\partial}{\partial x}\left(\frac{Q^2}{A}\right) + \frac{A}{\varrho}\frac{\partial P}{\partial x} = -2\pi(\gamma+2)\frac{\mu}{\varrho}\frac{Q}{A}, \tag{2.160}$$

where γ is a parameter related to the velocity profile, ϱ is the density of blood, and μ is the dynamic viscosity of blood.

We have three unknowns A, Q, and P, and two equations (2.159) and (2.160). A third equation is needed to relate the flow to the deformations of the wall. A common form for this equation is

$$\frac{\partial P}{\partial t} + \frac{1}{C}\frac{\partial Q}{\partial x} = 0, \tag{2.161}$$

where C is the compliance of the wall, given by the constitutive relation

$$C = \frac{\partial A}{\partial P} + \frac{\partial A}{\partial t}, \tag{2.162}$$

which requires a relationship between A and P. One common model is to view the vessel wall, locally, as a thin elastic tube subject to an internal pressure. This gives the relation

$$P = P_0 + \frac{\pi h E}{(1-\nu^2)A_0}\left(\sqrt{A} - \sqrt{A_0}\right),$$

where P_0 and A_0 are corresponding reference values when the wall is not deformed, h is the thickness of the wall, and E and ν are Young's modulus and Poisson's ratio of the elastic material in the wall. The derivative becomes

$$C = \frac{\partial A}{\partial P} = \frac{2(1-\nu^2)A_0}{\pi h E}\sqrt{A_0} + 2\left(\frac{(1-\nu^2)A_0}{\pi h E}\right)^2 (P - P_0)\,. \tag{2.163}$$

Another (nonlinear) deformation model of the wall, which has a better fit with experiments, is

$$P = P_0 \exp\left(\beta(A/A_0 - 1)\right),$$

where β is some parameter to be estimated. This law leads to

$$C = \frac{\partial A}{\partial P} = \frac{A_0}{\beta P}\,. \tag{2.164}$$

Reduction to the standard wave equation It is not uncommon to neglect the viscous term on the right-hand side of (2.160) and also the quadratic term with Q^2 on the left-hand side. The reduced equations (2.160) and (2.161) form a first-order linear wave equation system:

$$C\frac{\partial P}{\partial t} = -\frac{\partial Q}{\partial x}, \tag{2.165}$$

$$\frac{\partial Q}{\partial t} = -\frac{A}{\varrho}\frac{\partial P}{\partial x}. \tag{2.166}$$

These can be combined into standard 1D wave PDE by differentiating the first equation with respect to t and the second with respect to x,

$$\frac{\partial}{\partial t}\left(C\frac{\partial P}{\partial t}\right) = \frac{\partial}{\partial x}\left(\frac{A}{\varrho}\frac{\partial P}{\partial x}\right),$$

which can be approximated by

$$\frac{\partial^2 Q}{\partial t^2} = c^2\frac{\partial^2 Q}{\partial x^2}, \quad c = \sqrt{\frac{A}{\varrho C}}, \tag{2.167}$$

where the A and C in the expression for c are taken as constant reference values.

2.14.9 Electromagnetic Waves

Light and radio waves are governed by standard wave equations arising from Maxwell's general equations. When there are no charges and no currents, as in a vacuum, Maxwell's equations take the form

$$\begin{aligned}
\nabla\cdot\boldsymbol{E} &= 0,\\
\nabla\cdot\boldsymbol{B} &= 0,\\
\nabla\times\boldsymbol{E} &= -\frac{\partial\boldsymbol{B}}{\partial t},\\
\nabla\times\boldsymbol{B} &= \mu_0\epsilon_0\frac{\partial\boldsymbol{E}}{\partial t},
\end{aligned}$$

where $\epsilon_0 = 8.854187817620\cdot 10^{-12}$ (F/m) is the permittivity of free space, also known as the electric constant, and $\mu_0 = 1.2566370614\cdot 10^{-6}$ (H/m) is the permeability of free space, also known as the magnetic constant. Taking the curl of the two last equations and using the mathematical identity

$$\nabla\times(\nabla\times\boldsymbol{E}) = \nabla(\nabla\cdot\boldsymbol{E}) - \nabla^2\boldsymbol{E} = -\nabla^2\boldsymbol{E} \text{ when } \nabla\cdot\boldsymbol{E} = 0,$$

gives the wave equation governing the electric and magnetic field:

$$\frac{\partial^2\boldsymbol{E}}{\partial t^2} = c^2\nabla^2\boldsymbol{E}, \tag{2.168}$$

$$\frac{\partial^2\boldsymbol{B}}{\partial t^2} = c^2\nabla^2\boldsymbol{B}, \tag{2.169}$$

with $c = 1/\sqrt{\mu_0\epsilon_0}$ as the velocity of light. Each component of $\boldsymbol{E}$ and $\boldsymbol{B}$ fulfills a wave equation and can hence be solved independently.

2.15 Exercises

Exercise 2.20: Simulate waves on a non-homogeneous string
Simulate waves on a string that consists of two materials with different density. The tension in the string is constant, but the density has a jump at the middle of the string. Experiment with different sizes of the jump and produce animations that visualize the effect of the jump on the wave motion.

Hint According to Sect. 2.14.1, the density enters the mathematical model as ϱ in $\varrho u_{tt} = Tu_{xx}$, where T is the string tension. Modify, e.g., the `wave1D_u0v.py` code to incorporate the tension and two density values. Make a mesh function `rho` with density values at each spatial mesh point. A value for the tension may be 150 N. Corresponding density values can be computed from the wave velocity estimations in the `guitar` function in the `wave1D_u0v.py` file.
Filename: `wave1D_u0_sv_discont`.

Exercise 2.21: Simulate damped waves on a string
Formulate a mathematical model for damped waves on a string. Use data from Sect. 2.3.6, and tune the damping parameter so that the string is very close to the rest state after 15 s. Make a movie of the wave motion.
Filename: `wave1D_u0_sv_damping`.

Exercise 2.22: Simulate elastic waves in a rod
A hammer hits the end of an elastic rod. The exercise is to simulate the resulting wave motion using the model (2.128) from Sect. 2.14.2. Let the rod have length L and let the boundary $x = L$ be stress free so that $\sigma_{xx} = 0$, implying that $\partial u/\partial x = 0$. The left end $x = 0$ is subject to a strong stress pulse (the hammer), modeled as

$$\sigma_{xx}(t) = \begin{cases} S, & 0 < t \leq t_s, \\ 0, & t > t_s \end{cases} .$$

The corresponding condition on u becomes $u_x = S/E$ for $t \leq t_s$ and zero afterwards (recall that $\sigma_{xx} = Eu_x$). This is a non-homogeneous Neumann condition, and you will need to approximate this condition and combine it with the scheme (the ideas and manipulations follow closely the handling of a non-zero initial condition $u_t = V$ in wave PDEs or the corresponding second-order ODEs for vibrations).
Filename: `wave_rod`.

Exercise 2.23: Simulate spherical waves
Implement a model for spherically symmetric waves using the method described in Sect. 2.14.6. The boundary condition at $r = 0$ must be $\partial u/\partial r = 0$, while the condition at $r = R$ can either be $u = 0$ or a radiation condition as described in Problem 2.12. The $u = 0$ condition is sufficient if R is so large that the amplitude

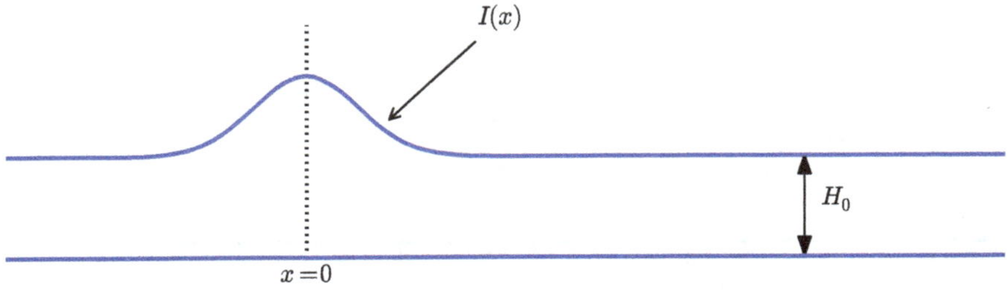

Fig. 2.11 Sketch of initial water surface due to a subsea earthquake

of the spherical wave has become insignificant. Make movie(s) of the case where the source term is located around $r = 0$ and sends out pulses

$$f(r,t) = \begin{cases} Q \exp\left(-\frac{r^2}{2\Delta r^2}\right) \sin \omega t, & \sin \omega t \geq 0 \\ 0, & \sin \omega t < 0 \end{cases}.$$

Here, Q and ω are constants to be chosen.

Hint Use the program `wave1D_u0v.py` as a starting point. Let `solver` compute the v function and then set $u = v/r$. However, $u = v/r$ for $r = 0$ requires special treatment. One possibility is to compute `u[1:] = v[1:]/r[1:]` and then set `u[0]=u[1]`. The latter makes it evident that $\partial u/\partial r = 0$ in a plot.
Filename: `wave1D_spherical`.

Problem 2.24: Earthquake-generated tsunami over a subsea hill
A subsea earthquake leads to an immediate lift of the water surface, see Fig. 2.11. The lifted water surface splits into two tsunamis, one traveling to the right and one to the left, as depicted in Fig. 2.12. Since tsunamis are normally very long waves, compared to the depth, with a small amplitude, compared to the wave length, a standard wave equation is relevant:

$$\eta_{tt} = (gH(x)\eta_x)_x,$$

where η is the elevation of the water surface, g is the acceleration of gravity, and $H(x)$ is the still water depth.

To simulate the right-going tsunami, we can impose a symmetry boundary at $x = 0$: $\partial\eta/\partial x = 0$. We then simulate the wave motion in $[0, L]$. Unless the ocean ends at $x = L$, the waves should travel undisturbed through the boundary $x = L$. A radiation condition as explained in Problem 2.12 can be used for this purpose. Alternatively, one can just stop the simulations before the wave hits the boundary at $x = L$. In that case it does not matter what kind of boundary condition we use at $x = L$. Imposing $\eta = 0$ and stopping the simulations when $|\eta_i^n| > \epsilon$, $i = N_x - 1$, is a possibility (ϵ is a small parameter).

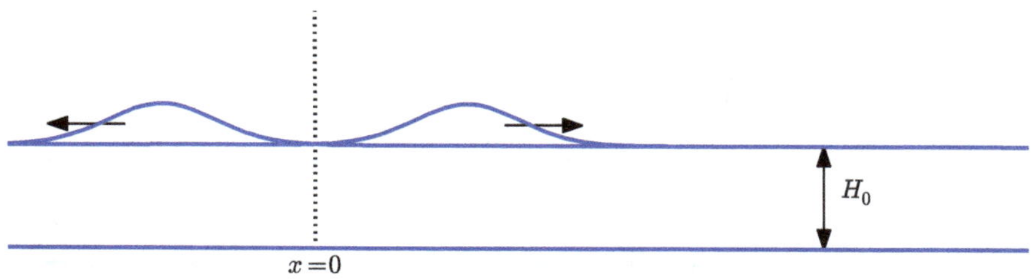

Fig. 2.12 An initial surface elevation is split into two waves

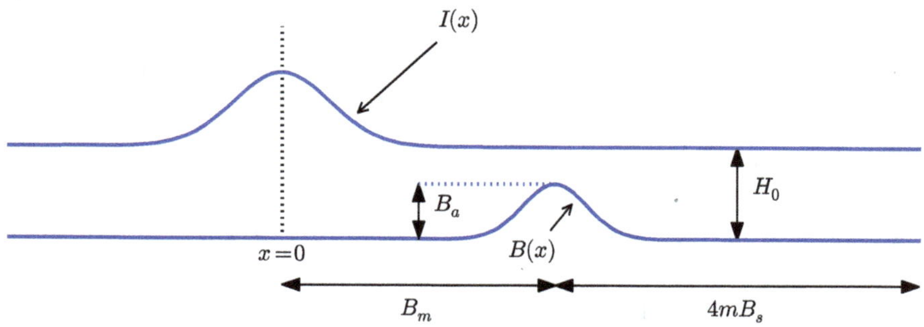

Fig. 2.13 Sketch of an earthquake-generated tsunami passing over a subsea hill

The shape of the initial surface can be taken as a Gaussian function,

$$I(x; I_0, I_a, I_m, I_s) = I_0 + I_a \exp\left(-\left(\frac{x - I_m}{I_s}\right)^2\right), \tag{2.170}$$

with $I_m = 0$ reflecting the location of the peak of $I(x)$ and I_s being a measure of the width of the function $I(x)$ (I_s is $\sqrt{2}$ times the standard deviation of the familiar normal distribution curve).

Now we extend the problem with a hill at the sea bottom, see Fig. 2.13. The wave speed $c = \sqrt{gH(x)} = \sqrt{g(H_0 - B(x))}$ will then be reduced in the shallow water above the hill.

One possible form of the hill is a Gaussian function,

$$B(x; B_0, B_a, B_m, B_s) = B_0 + B_a \exp\left(-\left(\frac{x - B_m}{B_s}\right)^2\right), \tag{2.171}$$

but many other shapes are also possible, e.g., a "cosine hat" where

$$B(x; B_0, B_a, B_m, B_s) = B_0 + B_a \cos\left(\pi \frac{x - B_m}{2B_s}\right), \tag{2.172}$$

when $x \in [B_m - B_s, B_m + B_s]$ while $B = B_0$ outside this interval.

Also an abrupt construction may be tried:

$$B(x; B_0, B_a, B_m, B_s) = B_0 + B_a, \tag{2.173}$$

for $x \in [B_m - B_s, B_m + B_s]$ while $B = B_0$ outside this interval.

The `wave1D_dn_vc.py` program can be used as starting point for the implementation. Visualize both the bottom topography and the water surface elevation in the same plot. Allow for a flexible choice of bottom shape: (2.171), (2.172), (2.173), or $B(x) = B_0$ (flat).

The purpose of this problem is to explore the quality of the numerical solution η_i^n for different shapes of the bottom obstruction. The "cosine hat" and the box-shaped hills have abrupt changes in the derivative of $H(x)$ and are more likely to generate numerical noise than the smooth Gaussian shape of the hill. Investigate if this is true.
Filename: `tsunami1D_hill`.

Problem 2.25: Earthquake-generated tsunami over a 3D hill
This problem extends Problem 2.24 to a three-dimensional wave phenomenon, governed by the 2D PDE

$$\eta_{tt} = (gH\eta_x)_x + (gH\eta_y)_y = \nabla \cdot (gH\nabla\eta)\,. \tag{2.174}$$

We assume that the earthquake arises from a fault along the line $x = 0$ in the xy-plane so that the initial lift of the surface can be taken as $I(x)$ in Problem 2.24. That is, a plane wave is propagating to the right, but will experience bending because of the bottom.

The bottom shape is now a function of x and y. An "elliptic" Gaussian function in two dimensions, with its peak at (B_{mx}, B_{my}), generalizes (2.171):

$$B = B_0 + B_a \exp\left(-\left(\frac{x - B_{mx}}{B_s}\right)^2 - \left(\frac{y - B_{my}}{bB_s}\right)^2\right), \tag{2.175}$$

where b is a scaling parameter: $b = 1$ gives a circular Gaussian function with circular contour lines, while $b \neq 1$ gives an elliptic shape with elliptic contour lines. To indicate the input parameters in the model, we may write

$$B = B(x; B_0, B_a, B_{mx}, B_{my}, B_s, b)\,.$$

The "cosine hat" (2.172) can also be generalized to

$$B = B_0 + B_a \cos\left(\pi\frac{x - B_{mx}}{2B_s}\right)\cos\left(\pi\frac{y - B_{my}}{2B_s}\right), \tag{2.176}$$

when $0 \le \sqrt{x^2 + y^2} \le B_s$ and $B = B_0$ outside this circle.

A box-shaped obstacle means that

$$B(x; B_0, B_a, B_m, B_s, b) = B_0 + B_a \tag{2.177}$$

for x and y inside a rectangle

$$B_{mx} - B_s \leq x \leq B_{mx} + B_s, \quad B_{my} - bB_s \leq y \leq B_{my} + bB_s,$$

and $B = B_0$ outside this rectangle. The b parameter controls the rectangular shape of the cross section of the box.

Note that the initial condition and the listed bottom shapes are symmetric around the line $y = B_{my}$. We therefore expect the surface elevation also to be symmetric with respect to this line. This means that we can halve the computational domain by working with $[0, L_x] \times [0, B_{my}]$. Along the upper boundary, $y = B_{my}$, we must impose the symmetry condition $\partial\eta/\partial n = 0$. Such a symmetry condition ($-\eta_x = 0$) is also needed at the $x = 0$ boundary because the initial condition has a symmetry here. At the lower boundary $y = 0$ we also set a Neumann condition (which becomes $-\eta_y = 0$). The wave motion is to be simulated until the wave hits the reflecting boundaries where $\partial\eta/\partial n = \eta_x = 0$ (one can also set $\eta = 0$ - the particular condition does not matter as long as the simulation is stopped before the wave is influenced by the boundary condition).

Visualize the surface elevation. Investigate how different hill shapes, different sizes of the water gap above the hill, and different resolutions $\Delta x = \Delta y = h$ and Δt influence the numerical quality of the solution.
Filename: `tsunami2D_hill`.

Problem 2.26: Investigate Mayavi for visualization
Play with Mayavi code for visualizing 2D solutions of the wave equation with variable wave velocity. See if there are effective ways to visualize both the solution and the wave velocity scalar field at the same time.
Filename: `tsunami2D_hill_mlab`.

Problem 2.27: Investigate visualization packages
Create some fancy 3D visualization of the water waves *and* the subsea hill in Problem 2.25. Try to make the hill transparent. Possible visualization tools are Mayavi[16], Paraview[17], and OpenDX[18].
Filename: `tsunami2D_hill_viz`.

Problem 2.28: Implement loops in compiled languages
Extend the program from Problem 2.25 such that the loops over mesh points, inside the time loop, are implemented in compiled languages. Consider implementations in Cython, Fortran via `f2py`, C via Cython, C via `f2py`, C/C++ via Instant, and C/C++ via `scipy.weave`. Perform efficiency experiments to investigate the relative performance of the various implementations. It is often advantageous to normalize CPU times by the fastest method on a given mesh.
Filename: `tsunami2D_hill_compiled`.

[16] http://code.enthought.com/projects/mayavi/
[17] http://www.paraview.org/
[18] http://www.opendx.org/

Exercise 2.29: Simulate seismic waves in 2D
The goal of this exercise is to simulate seismic waves using the PDE model (2.138) in a 2D xz domain with geological layers. Introduce m horizontal layers of thickness h_i, $i = 0, \ldots, m-1$. Inside layer number i we have a vertical wave velocity $c_{z,i}$ and a horizontal wave velocity $c_{h,i}$. Make a program for simulating such 2D waves. Test it on a case with 3 layers where

$$c_{z,0} = c_{z,1} = c_{z,2}, \quad c_{h,0} = c_{h,2}, \quad c_{h,1} \ll c_{h,0} .$$

Let s be a localized point source at the middle of the Earth's surface (the upper boundary) and investigate how the resulting wave travels through the medium. The source can be a localized Gaussian peak that oscillates in time for some time interval. Place the boundaries far enough from the expanding wave so that the boundary conditions do not disturb the wave. Then the type of boundary condition does not matter, except that we physically need to have $p = p_0$, where p_0 is the atmospheric pressure, at the upper boundary.
Filename: `seismic2D`.

Project 2.30: Model 3D acoustic waves in a room
The equation for sound waves in air is derived in Sect. 2.14.5 and reads

$$p_{tt} = c^2\nabla^2 p,$$

where $p(x, y, z, t)$ is the pressure and c is the speed of sound, taken as 340 m/s. However, sound is absorbed in the air due to relaxation of molecules in the gas. A model for simple relaxation, valid for gases consisting only of one type of molecules, is a term $c^2\tau_s\nabla^2 p_t$ in the PDE, where τ_s is the relaxation time. If we generate sound from, e.g., a loudspeaker in the room, this sound source must also be added to the governing equation.

The PDE with the mentioned type of damping and source then becomes

$$p_t t = c^2\nabla^p + c^2\tau_s\nabla^2 p_t + f, \tag{2.178}$$

where $f(x, y, z, t)$ is the source term.

The walls can absorb some sound. A possible model is to have a "wall layer" (thicker than the physical wall) outside the room where c is changed such that some of the wave energy is reflected and some is absorbed in the wall. The absorption of energy can be taken care of by adding a damping term bp_t in the equation:

$$p_t t + bp_t = c^2\nabla^p + c^2\tau_s\nabla^2 p_t + f\,. \tag{2.179}$$

Typically, $b = 0$ in the room and $b > 0$ in the wall. A discontinuity in b or c will give rise to reflections. It can be wise to use a constant c in the wall to control reflections because of the discontinuity between c in the air and in the wall, while b is gradually increased as we go into the wall to avoid reflections because of rapid changes in b. At the outer boundary of the wall the condition $p = 0$ or $\partial p/\partial n = 0$ can be imposed. The waves should anyway be approximately dampened to $p = 0$ this far out in the wall layer.

There are two strategies for discretizing the $\nabla^2 p_t$ term: using a center difference between times $n+1$ and $n-1$ (if the equation is sampled at level n), or use a one-sided difference based on levels n and $n-1$. The latter has the advantage of not leading to any equation system, while the former is second-order accurate as the scheme for the simple wave equation $p_t t = c^2\nabla^2 p$. To avoid an equation system, go for the one-sided difference such that the overall scheme becomes explicit and only of first order in time.

Develop a 3D solver for the specified PDE and introduce a wall layer. Test the solver with the method of manufactured solutions. Make some demonstrations where the wall reflects and absorbs the waves (reflection because of discontinuity in b and absorption because of growing b). Experiment with the impact of the τ_s parameter.
Filename: `acoustics`.

Project 2.31: Solve a 1D transport equation
We shall study the wave equation

$$u_t + cu_x = 0, \quad x \in (0, L],\ t \in (0, T], \tag{2.180}$$

with initial condition

$$u(x, 0) = I(x), \quad x \in [0, L], \tag{2.181}$$

and *one* periodic boundary condition

$$u(0, t) = u(L, t)\,. \tag{2.182}$$

This boundary condition means that what goes out of the domain at $x = L$ comes in at $x = 0$. Roughly speaking, we need only one boundary condition because the spatial derivative is of first order only.

Physical interpretation The parameter c can be constant or variable, $c = c(x)$. The equation (2.180) arises in *transport* problems where a quantity u, which could be temperature or concentration of some contaminant, is transported with the velocity c of a fluid. In addition to the transport imposed by "travelling with the fluid", u may also be transported by diffusion (such as heat conduction or Fickian diffusion), but we have in the model $u_t + cu_x$ assumed that diffusion effects are negligible, which they often are.

a) Show that under the assumption of $a = \text{const}$,

$$u(x, t) = I(x - ct) \tag{2.183}$$

fulfills the PDE as well as the initial and boundary condition (provided $I(0) = I(L)$).
A widely used numerical scheme for (2.180) applies a forward difference in time and a backward difference in space when $c > 0$:

$$[D_t^+ u + cD_x^- u = 0]_i^n\,. \tag{2.184}$$

For $c < 0$ we use a forward difference in space: $[cD_x^+ u]_i^n$.

b) Set up a computational algorithm and implement it in a function. Assume a is constant and positive.
c) Test the implementation by using the remarkable property that the numerical solution is exact at the mesh points if $\Delta t = c^{-1}\Delta x$.
d) Make a movie comparing the numerical and exact solution for the following two choices of initial conditions:

$$I(x) = \left[\sin\left(\pi\frac{x}{L}\right)\right]^{2n} \tag{2.185}$$

where n is an integer, typically $n = 5$, and

$$I(x) = \exp\left(-\frac{(x - L/2)^2}{2\sigma 2}\right). \tag{2.186}$$

Choose $\Delta t = c^{-1}\Delta x, 0.9c^{-1}\Delta x, 0.5c^{-1}\Delta x$.
e) The performance of the suggested numerical scheme can be investigated by analyzing the numerical dispersion relation. Analytically, we have that the *Fourier component*

$$u(x,t) = e^{i(kx-\omega t)},$$

is a solution of the PDE if $\omega = kc$. This is the *analytical dispersion relation*. A complete solution of the PDE can be built by adding up such Fourier components with different amplitudes, where the initial condition I determines the amplitudes. The solution u is then represented by a Fourier series.
A similar discrete Fourier component at (x_p, t_n) is

$$u_p^q = e^{i(kp\Delta x - \tilde{\omega} n\Delta t)},$$

where in general $\tilde{\omega}$ is a function of k, Δt, and Δx, and differs from the exact $\omega = kc$.
Insert the discrete Fourier component in the numerical scheme and derive an expression for $\tilde{\omega}$, i.e., the discrete dispersion relation. Show in particular that if $\Delta t/(c\Delta x) = 1$, the discrete solution coincides with the exact solution at the mesh points, regardless of the mesh resolution (!). Show that if the stability condition

$$\frac{\Delta t}{c\Delta x} \le 1,$$

the discrete Fourier component cannot grow (i.e., $\tilde{\omega}$ is real).
f) Write a test for your implementation where you try to use information from the numerical dispersion relation.
We shall hereafter assume that $c(x) > 0$.
g) Set up a computational algorithm for the variable coefficient case and implement it in a function. Make a test that the function works for constant a.
h) It can be shown that for an observer moving with velocity $c(x)$, u is constant. This can be used to derive an exact solution when a varies with x. Show first that

$$u(x,t) = f(C(x) - t), \tag{2.187}$$

where

$$C'(x) = \frac{1}{c(x)},$$

is a solution of (2.180) for any differentiable function f.

i) Use the initial condition to show that an exact solution is

$$u(x,t) = I(C^{-1}(C(x)-t)),$$

with C^{-1} being the inverse function of $C = \int c^{1}dx$. Since $C(x)$ is an integral $\int_0^x (1/c)dx$, $C(x)$ is monotonically increasing and there exists hence an inverse function C^{-1} with values in $[0, L]$.

To compute (2.187) we need to integrate $1/c$ to obtain C and then compute the inverse of C.

The inverse function computation can be easily done if we first think discretely. Say we have some function $y = g(x)$ and seek its inverse. Plotting (x_i, y_i), where $y_i = g(x_i)$ for some mesh points x_i, displays g as a function of x. The inverse function is simply x as a function of g, i.e., the curve with points (y_i, x_i). We can therefore quickly compute points at the curve of the inverse function. One way of extending these points to a continuous function is to assume a linear variation (known as linear interpolation) between the points (which actually means to draw straight lines between the points, exactly as done by a plotting program).

The function `wrap2callable` in `scitools.std` can take a set of points and return a continuous function that corresponds to linear variation between the points. The computation of the inverse of a function g on $[0, L]$ can then be done by

```
def inverse(g, domain, resolution=101):
    x = linspace(domain[0], domain[L], resolution)
    y = g(x)
    from scitools.std import wrap2callable
    g_inverse = wrap2callable((y, x))
    return g_inverse
```

To compute $C(x)$ we need to integrate $1/c$, which can be done by a Trapezoidal rule. Suppose we have computed $C(x_i)$ and need to compute $C(x_{i+1})$. Using the Trapezoidal rule with m subintervals over the integration domain $[x_i, x_{i+1}]$ gives

$$C(x_{i+1}) = C(x_i) + \int_{x_i}^{x_{i+1}} \frac{dx}{c} \approx h\left(\frac{1}{2}\frac{1}{c(x_i)} + \frac{1}{2}\frac{1}{c(x_{i+1})} + \sum_{j=1}^{m-1} \frac{1}{c(x_i + jh)}\right), \tag{2.188}$$

where $h = (x_{i+1} - x_i)/m$ is the length of the subintervals used for the integral over $[x_i, x_{i+1}]$. We observe that (2.188) is a *difference equation* which we can solve by repeatedly applying (2.188) for $i = 0, 1, \ldots, N_x - 1$ if a mesh $x_0, x, \ldots, x_{N_x}$ is prescribed. Note that $C(0) = 0$.

j) Implement a function for computing $C(x_i)$ and one for computing $C^{-1}(x)$ for any x. Use these two functions for computing the exact solution $I(C^{-1}(C(x) - t))$. End up with a function `u_exact_variable_c(x, n, c, I)` that returns the value of $I(C^{-1}(C(x) - t_n))$.
k) Make movies showing a comparison of the numerical and exact solutions for the two initial conditions (2.185) and (2.15). Choose $\Delta t = \Delta x / \max_{0,L} c(x)$ and the velocity of the medium as
(a) $c(x) = 1 + \epsilon \sin(k\pi x/L)$, $\epsilon < 1$,
(b) $c(x) = 1 + I(x)$, where I is given by (2.185) or (2.15).
The PDE $u_t + cu_x = 0$ expresses that the initial condition $I(x)$ is transported with velocity $c(x)$.

Filename: `advec1D`.

Problem 2.32: General analytical solution of a 1D damped wave equation
We consider an initial-boundary value problem for the damped wave equation:

$$
\begin{aligned}
u_{tt} + bu_t &= c^2 u_{xx}, \quad x \in (0, L),\ t \in (0, T] \\
u(0, t) &= 0, \\
u(L, t) &= 0, \\
u(x, 0) &= I(x), \\
u_t(x, 0) &= V(x) .
\end{aligned}
$$

Here, $b \geq 0$ and c are given constants. The aim is to derive a general analytical solution of this problem. Familiarity with the method of separation of variables for solving PDEs will be assumed.

a) Seek a solution on the form $u(x, t) = X(x)T(t)$. Insert this solution in the PDE and show that it leads to two differential equations for X and T:

$$T'' + bT' + \lambda T = 0, \quad c^2 X'' + \lambda X = 0,$$

with $X(0) = X(L) = 0$ as boundary conditions, and λ as a constant to be determined.
b) Show that $X(x)$ is on the form

$$X_n(x) = C_n \sin kx, \quad k = \frac{n\pi}{L}, \quad n = 1, 2, \ldots$$

where C_n is an arbitrary constant.
c) Under the assumption that $(b/2)^2 < k^2$, show that $T(t)$ is on the form

$$T_n(t) = e^{-\frac{1}{2}bt}(a_n \cos \omega t + b_n \sin \omega t), \quad \omega = \sqrt{k^2 - \frac{1}{4}b^2}, \quad n = 1, 2, \ldots$$

The complete solution is then

$$u(x, t) = \sum_{n=1}^{\infty} \sin kx e^{-\frac{1}{2}bt}(A_n \cos \omega t + B_n \sin \omega t),$$

where the constants A_n and B_n must be computed from the initial conditions.

d) Derive a formula for A_n from $u(x, 0) = I(x)$ and developing $I(x)$ as a sine Fourier series on $[0, L]$.
e) Derive a formula for B_n from $u_t(x, 0) = V(x)$ and developing $V(x)$ as a sine Fourier series on $[0, L]$.
f) Calculate A_n and B_n from vibrations of a string where $V(x) = 0$ and

$$I(x) = \begin{cases} ax/x_0, & x < x_0, \\ a(L-x)/(L-x_0), & \text{otherwise}. \end{cases} \tag{2.189}$$

g) Implement a function `u_series(x, t, tol=1E-10)` for the series for $u(x, t)$, where `tol` is a tolerance for truncating the series. Simply sum the terms until $|a_n|$ and $|b_b|$ both are less than `tol`.
h) What will change in the derivation of the analytical solution if we have $u_x(0, t) = u_x(L, t) = 0$ as boundary conditions? And how will you solve the problem with $u(0, t) = 0$ and $u_x(L, t) = 0$?

Filename: `damped_wave1D`.

Problem 2.33: General analytical solution of a 2D damped wave equation
Carry out Problem 2.32 in the 2D case: $u_{tt} + bu_t = c^2(u_{xx} + u_{yy})$, where $(x, y) \in (0, L_x) \times (0, L_y)$. Assume a solution on the form $u(x, y, t) = X(x)Y(y)T(t)$.
Filename: `damped_wave2D`.

Permissions

All chapters in this book were first published in FDCPDES, by Springer by Hans Petter Langtangen and Svein Linge ; hereby published with permission under the Creative Commons Attribution License or equivalent. Every chapter published in this book has been scrutinized by our experts. Their significance has been extensively debated. The topics covered herein carry significant information for a comprehensive understanding. They may even be implemented as practical applications or may be referred to as a beginning point for further studies.

We would like to thank the editorial team for lending their expertise to make the book truly unique. They have played a crucial role in the development of this book. Without their invaluable contributions this book wouldn't have been possible. They have made vital efforts to compile up to date information on the varied aspects of this subject to make this book a valuable addition to the collection of many professionals and students.

This book was conceptualized with the vision of imparting up-to-date and integrated information in this field. To ensure the same, a matchless editorial board was set up. Every individual on the board went through rigorous rounds of assessment to prove their worth. After which they invested a large part of their time researching and compiling the most relevant data for our readers.

The editorial board has been involved in producing this book since its inception. They have spent rigorous hours researching and exploring the diverse topics which have resulted in the successful publishing of this book. They have passed on their knowledge of decades through this book. To expedite this challenging task, the publisher supported the team at every step. A small team of assistant editors was also appointed to further simplify the editing procedure and attain best results for the readers.

Apart from the editorial board, the designing team has also invested a significant amount of their time in understanding the subject and creating the most relevant covers. They scrutinized every image to scout for the most suitable representation of the subject and create an appropriate cover for the book.

The publishing team has been an ardent support to the editorial, designing and production team. Their endless efforts to recruit the best for this project, has resulted in the accomplishment of this book. They are a veteran in the field of academics and their pool of knowledge is as vast as their experience in printing. Their expertise and guidance has proved useful at every step. Their uncompromising quality standards have made this book an exceptional effort. Their encouragement from time to time has been an inspiration for everyone.

The publisher and the editorial board hope that this book will prove to be a valuable piece of knowledge for students, practitioners and scholars across the globe.

Index

www.ingramcontent.com/pod-product-compliance
Lightning Source LLC
LaVergne TN
LVHW080603210926
848485LV00003B/97

* 9 7 8 1 6 3 9 8 9 2 0 3 7 *